Guide for the Total Solar Eclipse of 2028

Color Edition

Fred Espenak

Edition 1.0
August 2024

Guide for the Total Solar Eclipse of 2028 – Color Edition

Astropixels Publishing
P.O. Box 16197
Portal, AZ 85632

Astropixels Publishing Website: *www.astropixels.com/pubs*

This book may be ordered at: *www.astropixels.com/pubs/Guide2028.html*

Printed in the United States of America

ISBN 978-1-941983-48-5

Astropixels Publication: AP046 (Version 1.0c)

First Edition

Front Cover: 1) A map of the path of the 2028 total solar eclipse through Australia and New Zealand.
2) Background: Sydney Opera House at night. ©2023, F. Espenak.

Back Cover: 1) Pat and Fred Espenak with the Sydney Harbour Bridge ©2023, F. Espenak.

More about the eclipse can be found at:

www.eclipsewise.com/solar/SEnews/TSE2028.html

Table of Contents

Photo 1 – The Sun's corona is only visible to the naked eye for a few minutes during a total eclipse of the Sun. Total solar eclipse of August 21, 2017 from Casper, Wyoming. © 2017 F. Espenak, www.MrEclipse.com

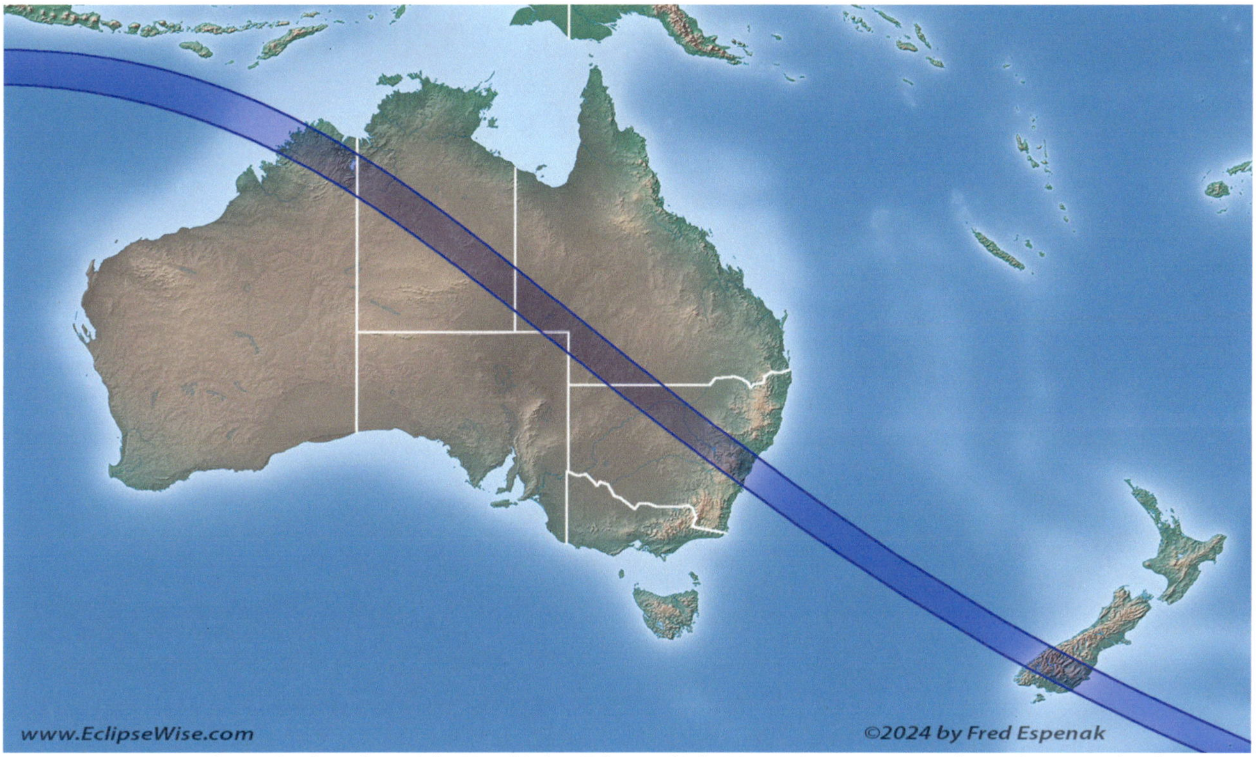

Figure 1 – The path of totality of the Total Solar Eclipse of July 22, 2028 crosses Australia and New Zealand.

1 – Introduction

On Saturday, July 22, 2028, a total eclipse of the Sun will be visible from the Eastern Hemisphere. Although a partial eclipse will be seen from South Asia, the Philippines, Papua New Guinea, and Indonesia, the total phase in which the Moon completely covers the Sun (known as totality) will only be seen from within the narrow path of the Moon's umbral shadow as it sweeps across parts of Australia and New Zealand (Figure 1).

During totality, the Sun's faint and exquisitely beautiful outer atmosphere – the solar corona – can be seen with the naked eye, the landscape is plunged into an eerie twilight, and brighter stars and planets appear. A total solar eclipse is one of the most awesome natural spectacles, but to witness it you must be in the *path of totality*.

The path of the Moon's umbral shadow begins in the Indian Ocean and travels east to reach the Cocos Islands, Christmas Island, and Western Australia (WA). Greatest eclipse occurs in WA where the central line duration of totality lasts 5 minutes 10 seconds. The track continues through remote parts of Australia's Northern Territory and South Australia before reaching parts of the more populated states of Queensland and New South Wales. Sydney (population 5.3 million) is the largest Australian city in the path where totality lasts 3 minutes 48 seconds.

Crossing the Tasman Sea, the Moon's umbral track reaches New Zealand's South Island where the cities of Queenstown and Dunedin lie in the path of totality. The central duration in both cities is just under 3 minutes.

The eclipse track ends 900 kilometers southeast of New Zealand's South Island in the South Pacific Ocean.

A partial eclipse will be visible from a much larger region including most of Australia, New Zealand, South Asia, the Philippines, Papua New Guinea, and Indonesia. Eclipse glasses or some type of eye protection must be used for viewing the partial eclipse. Of course, those in the path of the total eclipse must remove protective glasses to witness totality itself.

This guide is designed to assist observers in locating and reaching the central eclipse path to enjoy the awesome experience of totality.

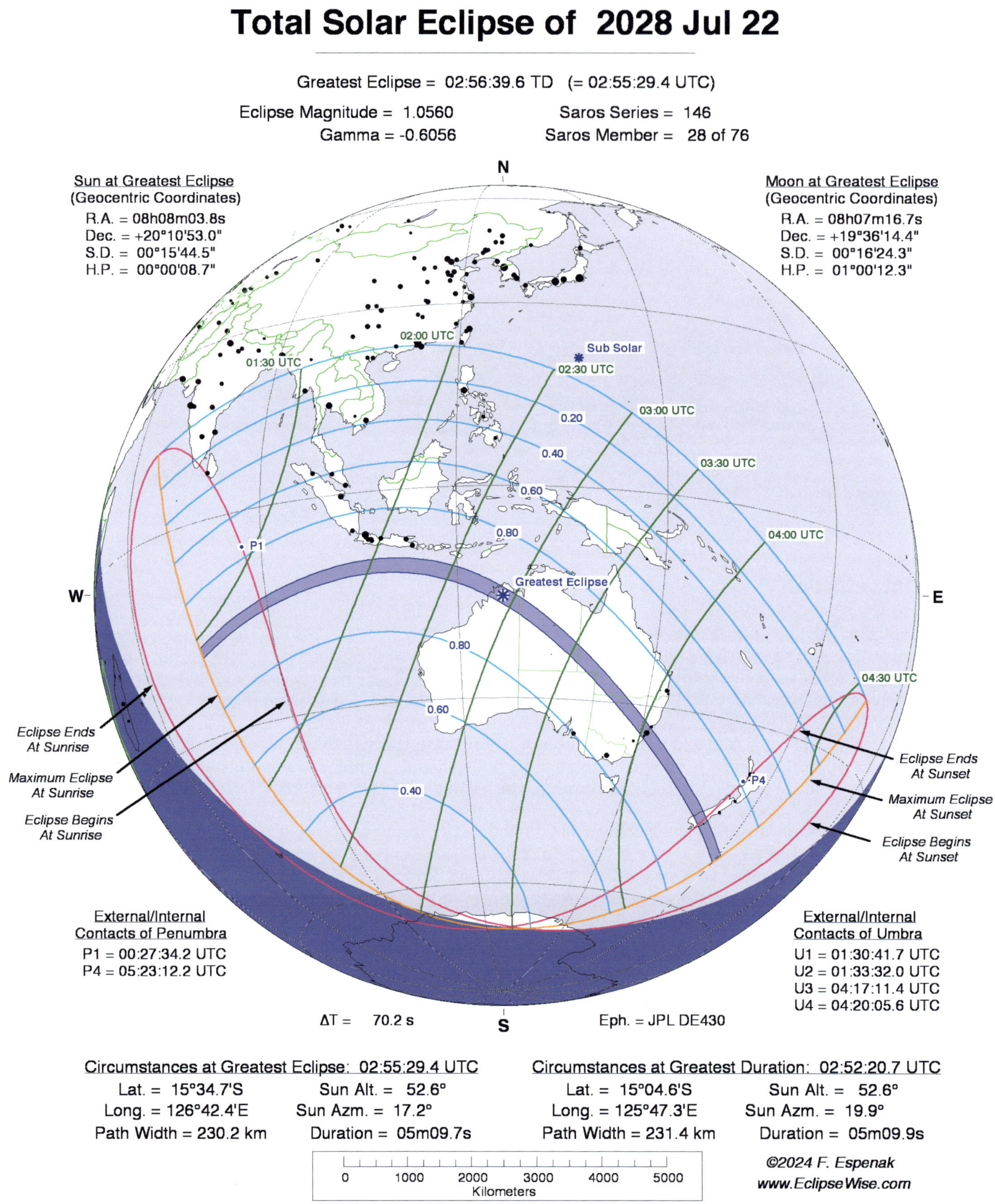

Figure 2 – Orthographic projection map of the Total Solar Eclipse of 2028.
Based on map from ***21st Century Canon of Solar Eclipses*** *by Fred Espenak (2017)*

2 – Orthographic Projection Map of the 2028 Total Solar Eclipse Path

The orthographic projection map of Earth in *Figure 2* shows the path of penumbral (partial eclipse) and umbral (total eclipse) shadows of the Moon during the eclipse.

The limits of the Moon's penumbral shadow define the region of visibility of the partial eclipse. It covers about half of Earth's daylight hemisphere and consists of several distinct zones or limits. At the northern and southern boundaries lie the limits of the penumbra's path. Great loops at the western and eastern extremes of the penumbra's path identify the areas where the eclipse begins and ends at sunrise and sunset, respectively.

Bisecting the *eclipse begins and ends at sunrise and sunset* loops is the curve of maximum eclipse at sunrise (western loop) and sunset (eastern loop). Points *P1* and *P4* mark the coordinates where the penumbral shadow first contacts (partial eclipse begins) and last contacts (partial eclipse ends) Earth's surface. The path of the umbral shadow (*path of totality*) appears within the penumbral path. Curves of constant eclipse magnitude define points where the maximum magnitude has values of 0.2, 0.4, 0.6, and 0.8.

Above the map is the instant of greatest eclipse[1], in Terrestrial Time[2] (TT) and Coordinated Universal Time[3] (UTC). The *Eclipse Magnitude* is the geocentric ratio of diameters of the Moon and the Sun. *Gamma* is the minimum distance of the Moon's shadow axis from Earth's center in Earth radii at greatest eclipse. Finally, the *Saros Series* of the eclipse is listed, followed by the *Saros Member* that gives the sequence position of the eclipse in the Saros, and the total number of eclipses in the series

In the upper left and right corners are the geocentric coordinates of the Sun and the Moon, respectively, at the instant of greatest eclipse. They are:

- R.A. – Right Ascension
- Dec. – Declination
- S.D. – Apparent Semi-Diameter
- H.P. – Horizontal Parallax

To the lower left are the contact times of the Moon's penumbral shadow with Earth. They mark the times when the partial eclipse begins and ends, globally. To the lower right are the contact times of the Moon's umbral shadow with Earth. They mark the times when the total eclipse path begins and ends, globally.

At bottom left are the geographic coordinates of the position of greatest eclipse along with the local circumstances at that location (i.e., Sun altitude, Sun azimuth, path width and duration of totality).

At bottom right are the corresponding values for the position of greatest duration[4]. Although the geographic locations of greatest duration and greatest eclipse differ by about 113 km (70 miles) along the eclipse track, the difference in the duration of totality is 0.18 second.

3 – Overview Maps of the Path of Totality

The overview map in *Figure 3* shows the path of the total eclipse through Australia and New Zealand.

Figure 4 features a close-up of the path through Australia while *Figure 5* focuses on the path through New Zealand. The white lines running across the eclipse path mark the position of mid-eclipse at 5-minute intervals. The positions of *Greatest Eclipse* and *Greatest Duration* (both in Western Australia) are marked *GE* and *GD*, respectively.

The Mercator projection used in this map preserves local directions and shapes at the expense of inflating the size of objects the further they are from the equator.

---

1 The instant of greatest eclipse occurs when the Moon's shadow axis passes closest to Earth's center.

2 Terrestrial Time or TT is a time scale established by astronomers to calculate the location or planetary position of other objects as they relate to the center of Earth and its rotation.

3 Coordinated Universal Time or UTC is the primary time standard that regulates world clocks and time. It is based on International Atomic Time (TAI) with leap seconds added at irregular intervals to compensate for the slowing of Earth's rotation. UTC is not adjusted for daylight saving time. It is effectively a successor to Greenwich Mean Time (GMT).

4 Greatest duration is the time (and location) when the length of totality reaches its maximum value.

Overview Map: Path of Totality – Australia and New Zealand

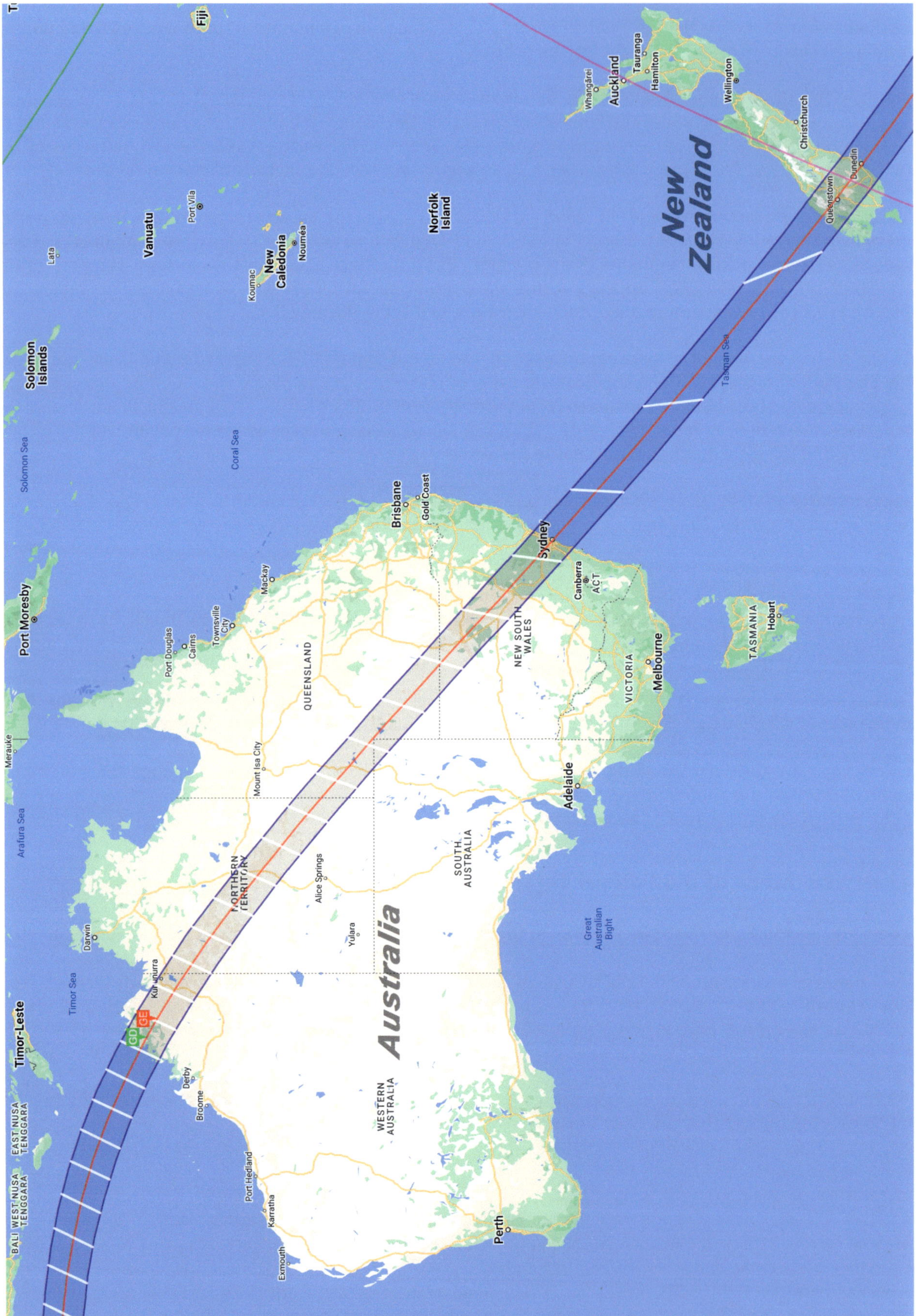

Figure 3 – Overview Map: Path of Totality – Australia and New Zealand

Overview Map: Path of Totality – Australia

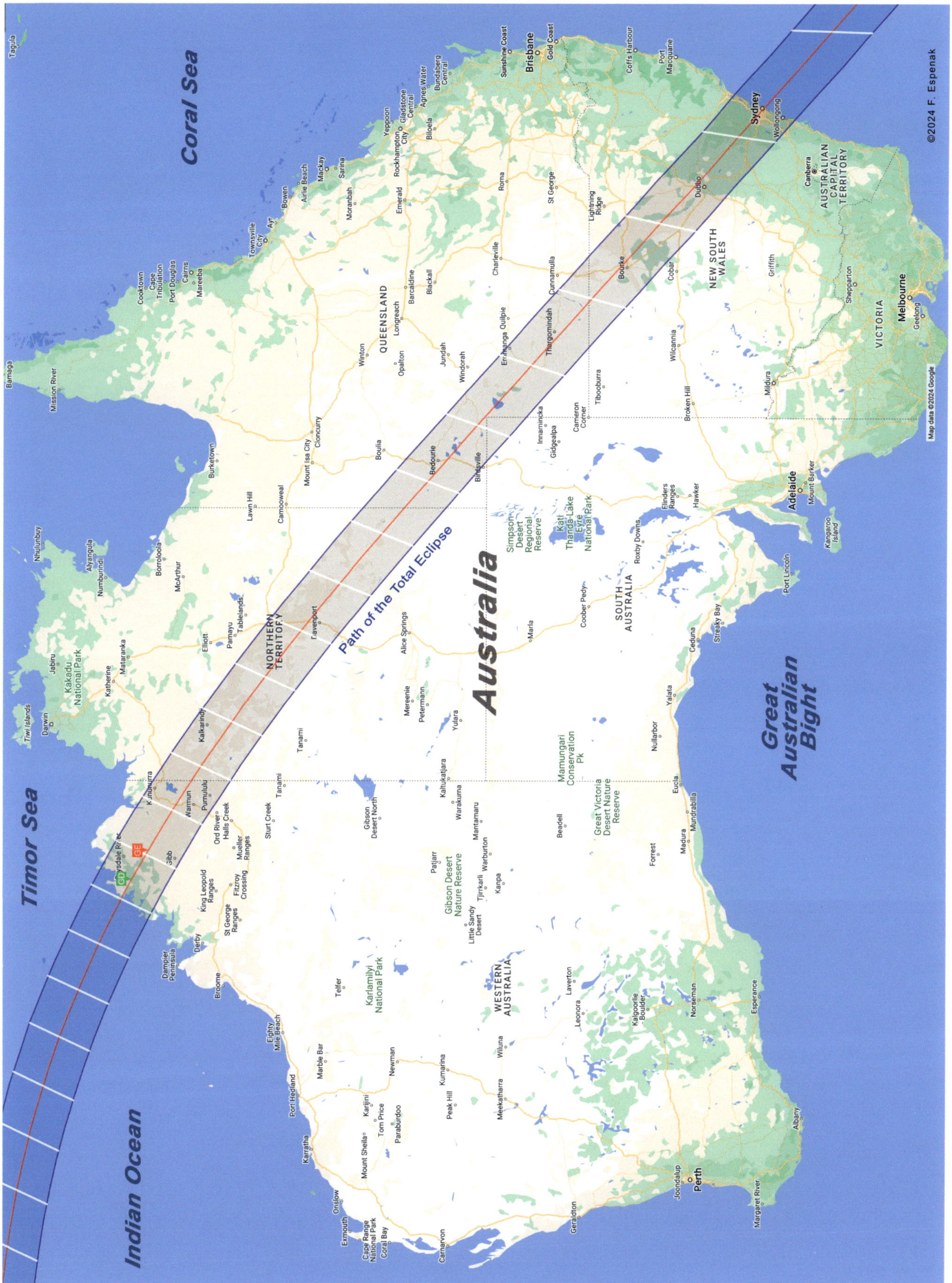

Figure 4 – Overview Map: Path of Totality – Australia

Overview Map: Path of Totality – New Zealand

Figure 5 – Overview Map: Path of Totality – New Zealand

4 – Path of Totality

The path of the Moon's umbral shadow begins at 01:32 UTC in the central Indian Ocean about 2900 km (1800 mi) south of India. Along the sunrise terminator, the duration of totality is 2m 28s (i.e., 2 minutes 28 seconds) as seen from the center of the narrow, 157 km (98 mi) wide path. No landfall occurs during the first 10 minutes of the central eclipse as the shadow sweeps northeast across the Indian Ocean.

Table 1 – Time Zones used by Countries within the 2028 Path of Totality

Time Zone Name	Abbrev.	Offset	Locations in 2028 Path
Cocos Islands Time	CCT	UTC + 6.5 hrs	Cocos Islands
Christmas Island Time	CXT	UTC + 7 hrs	Christmas Island
Australian Western Standard Time	AWST	UTC + 8 hrs	AUS: Western Australia
Australian Central Standard Time	ACST	UTC + 9.5 hrs	AUS: Northern Territory
Australian Eastern Standard Time	AEST	UTC +10 hrs	AUS: Queensland & New South Wales
New Zealand Standard Time	NZST	UTC + 12 hrs	New Zealand

Figure 6 – The path of totality crosses the Indian Ocean and encounters the Cocos (Keeling) Islands and Christmas Island.

The umbra finally reaches land at the Cocos (Keeling) Islands (Figure 6 and Map 01) at 01:42 UTC (08:12 am CCT). The Sun's altitude is 24° and the track width is 188 km (116 mi). West Island (in the Cocos) lies 21 km (16mi) south of the central line and the duration of totality is 3m 27s.

Continuing eastward, the next landfall of the Moon's umbral shadow is Christmas Island (Figure 6 and Map 02) at 01:55 UTC (08:55 am CXT). Located 36 km (22 mi) north of the central line, Fishing Fly Cove experiences totality lasting 3m 53s with the Sun 35° above the horizon.

The path of totality gradually curves to the southeast and reaches the northwestern coast of Australia in the Kimberley region[5] (Figure 7 and Map 03) at 02:50 UTC (10:50 am AWST). The umbra's ground velocity is 0.60 km/s and has nearly reached the midpoint of its long track across Earth. Shortly after entering Australia, the eclipse reaches the instant of Greatest Eclipse at 02:55:30 UTC (10:55:30 am AWST) with a duration of 5m 9.72s and the Sun's altitude at 52.6°. The point of greatest duration occurs three minutes earlier and about 113 km (70 miles) to the northwest. Greatest duration is 5m 9.90s, only 0.18s more than greatest eclipse.

---

[5] With a population less than 40,000 and spread over an area the size California, the Kimberley region in Western Australia's north-western corner is one of the world's last wilderness frontiers.

Figure 7 – The path of totality through Western Australia and Northern Territory.

Good weather prospects and a long duration of totality both favor the path through Western Australia and the Northwestern Territory. Unfortunately, these regions remain some of the most remote in all of Australia. Nevertheless, the determined eclipse observer may find easier access to the path as it passes 210 km (133 mi) north of Alice Springs (Figure 8). The central line duration of totality in this region is just under 5 minutes.

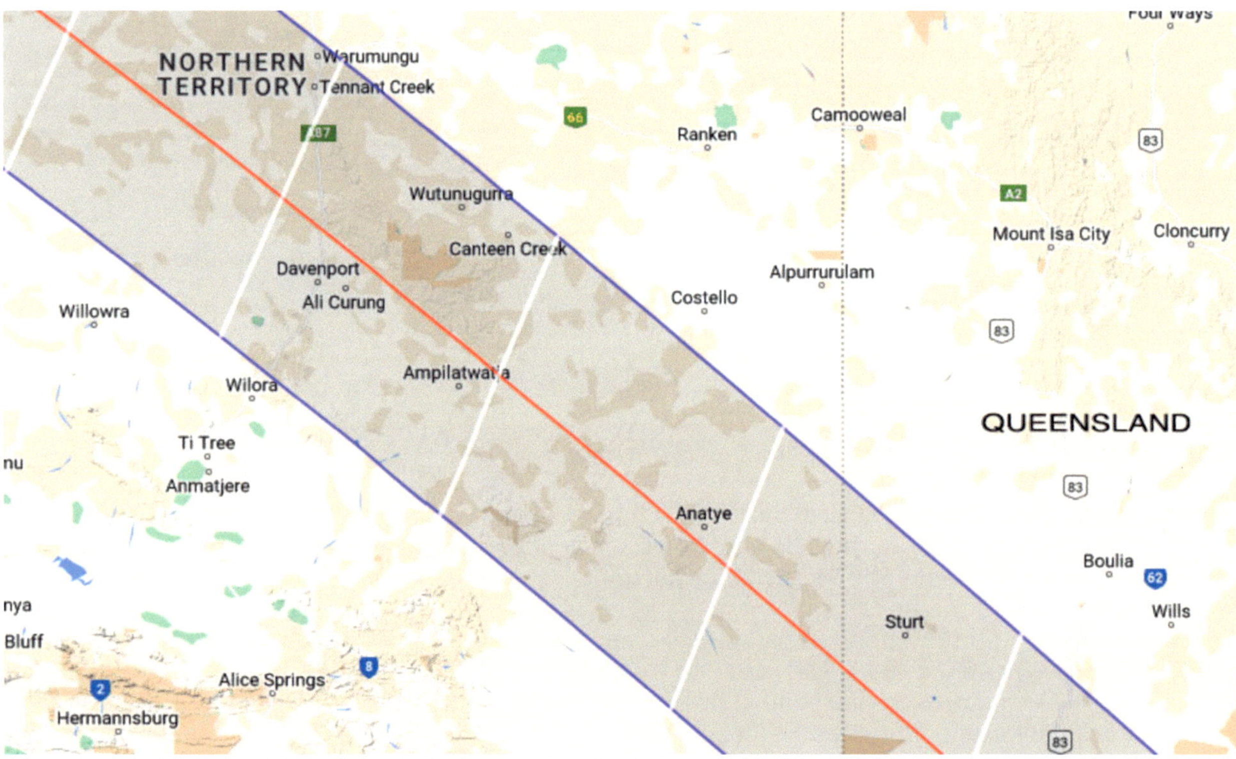

Figure 8 – The path of totality in central Australia through Northern Territory and Queensland.

At 03:32 UTC (1:32 pm AEST), the path enters Queensland. The central duration has decreased to 4m 43s and the Sun's altitude is 45°. Just clipping the northeastern corner of South Australia, the path continues through Queensland and reaches the state's southern border with New South Wales (NSW) at 03:50 UTC (1:50 pm AEST). The central line duration is 4m 14s as the Sun stands 37° above the horizon.

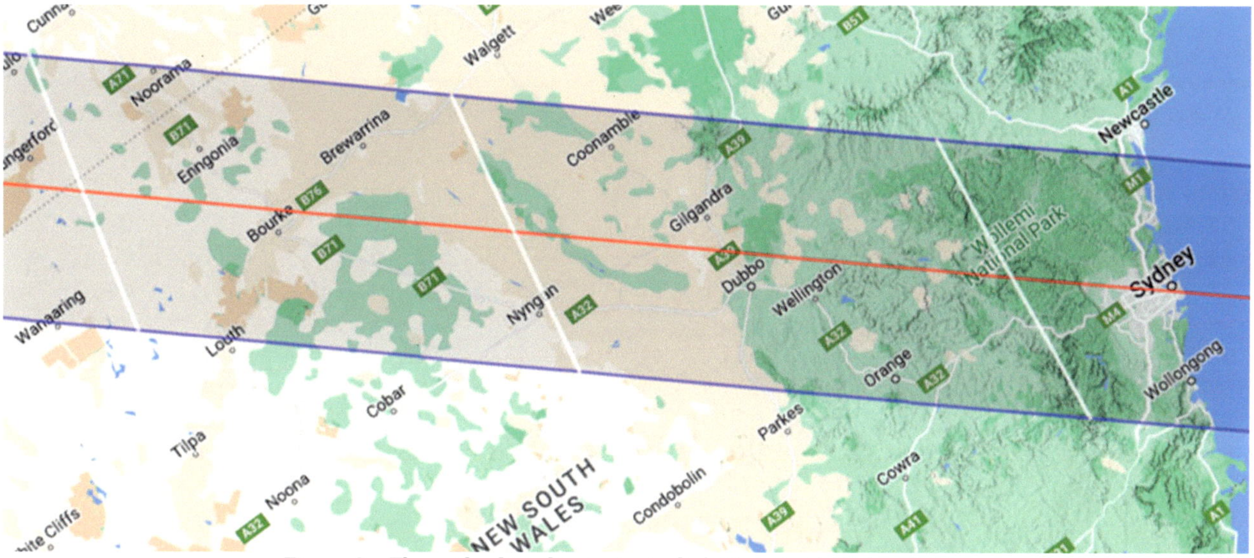

Figure 9 – The path of totality in Australia's New South Wales (NSW).

Bourke is one of the first sizable towns in the path of totality. With a population of about 2000, its citizens will witness a totality lasting 4m 6s. Nyngan, another similarly sized town in NSW is further from the central line but still gets 3m 15s of totality.

By 04:00 UTC (2:00 pm AEST), the umbra reaches Australia's Blue Mountains National Park, a World Heritage site and home of the famous Three Sisters in Katoomba. The central line duration is 3m 52s and the Sun is 30°. Less than two minutes later, the Moon's umbral shadow encounters the largest city in the 2028 eclipse path.

Sydney, the capital city of New South Wales and the second most populous city in Australia (population 5.4 million) is on the central line where totality lasts 3m 44s. Since the width of the path of totality is 187 km (116 mi), virtually all the suburbs surrounding Sydney also experience the total eclipse.

Leaving Australia and traveling with a ground velocity of 1.43 km/s (3,200 mi/hr), the lunar shadow spends the next 15 minutes crossing the Tasman Sea.

At 04:16 UTC (4:16 pm NZST), the track reaches the western coast of New Zealand's South Island (Figure 5). Milford Sound, with its towering peaks, cascading waterfalls, lies in the center of the path that is 171 km (106 mi) wide. Crossing the rugged Southern Alps, the shadow reaches Queenstown situated on the northwestern edge of Lake Wakatipu. The duration of totality is 2m 55s from the resort town, and the Sun's altitude is 10°.

Continuing to the east across the broad Central Otago Plains, the shadow reaches Dunedin (pop. 135,000), the second-largest city in the South Island. The duration of totality 2m 51s.

Leaving New Zealand at 04:17 UTC (4:17 pm NZST) with a velocity of 4.8 km/s (10,700 mi/hr) that is rapidly increasing, the umbra races 900 km (550 mi) across the South Pacific Ocean before reaching the end of the track at 04:18:39 UTC and returning to space. Over a period of 2 hours 47 minutes, the umbra travels along a 11,978 km (7443 mi) long track that covers an area equal to 0.46% of Earth's surface.

Table 2 – The 2028 Path of Totality

UTC	Latitude	Longitude	Diameter Ratio	Sun Alt	Sun Azm	Path Width km	Major Axis km	Minor Axis km	Duration Totality
01:32.1	18°23.4'S	075°36.9'E	1.041	0.0°	68.7°	157.4	—	137.2	02m28.3s
01:40	12°42.5'S	093°27.4'E	1.047	19.7°	62.8°	186.2	464.5	156.9	03m21.2s
01:50	11°10.3'S	101°29.6'E	1.050	29.5°	59.0°	202.9	337.2	166.1	03m52.8s
02:00	10°43.2'S	107°06.5'E	1.052	36.4°	54.8°	215.4	289.9	172.1	04m16.7s
02:10	10°50.7'S	111°36.7'E	1.053	41.8°	50.0°	224.6	264.7	176.5	04m35.5s
02:20	11°22.3'S	115°28.1'E	1.054	46.0°	44.4°	230.8	249.7	179.8	04m50.1s
02:30	12°13.1'S	118°54.4'E	1.055	49.2°	37.7°	233.9	240.4	182.1	05m00.5s
02:40	13°20.3'S	122°04.6'E	1.056	51.4°	30.2°	234.2	235.0	183.7	05m07.1s
02:50	14°42.9'S	125°04.7'E	1.056	52.5°	22.0°	232.1	232.5	184.6	05m09.8s
03:00	16°20.4'S	127°59.9'E	1.056	52.5°	13.4°	228.3	232.6	184.8	05m08.9s
03:10	18°13.2'S	130°55.1'E	1.056	51.5°	4.8°	223.2	235.3	184.3	05m04.4s
03:20	20°22.5'S	133°55.6'E	1.056	49.4°	356.7°	217.4	241.0	183.2	04m56.6s
03:30	22°50.8'S	137°08.2'E	1.055	46.3°	349.1°	211.1	250.4	181.3	04m45.6s
03:40	25°42.0'S	140°42.2'E	1.054	42.1°	342.1°	204.4	265.5	178.5	04m31.5s
03:50	29°03.7'S	144°53.4'E	1.053	36.8°	335.4°	197.3	290.3	174.5	04m13.9s
04:00	33°11.0'S	150°13.2'E	1.051	30.1°	328.7°	189.5	336.2	169.0	03m52.3s
04:10	38°45.7'S	158°09.9'E	1.048	20.6°	320.8°	179.8	453.0	160.6	03m24.1s
04:18.7	50°30.2'S	179°30.8'W	1.042	0.0°	302.8°	161.3	—	140.2	02m31.6s

Table 2 lists the geographic coordinates of the central line for every ten minutes UTC. *Diameter Ratio* gives the Moon's apparent diameter divided by the Sun's. The Sun's altitude (*Alt*) and azimuth (*Azm*) are given along with the width of the path of totality (kilometers). The *Major* and *Minor* axes of the umbral shadow are also listed. Finally, the *Duration of Totality* is given (minutes and seconds).

5 – Local Circumstances for the Eclipse

Local circumstances for the ***Total Eclipse*** for 59 cities are found in Table 3 (Australia) and Table 4 (New Zealand). All times are given in local time (see Table 1 for Time Zones). The *Duration Totality* is the period between the beginning and end of totality. These times include the effects of the Moon's limb profile. The Sun's altitude is given at the instant of mid-totality.

Local circumstances for the ***Partial Eclipse*** for over 65 cities are found in Table 5 (Australia), Table 6 (New Zealand), and Table 7 (Asia, Indonesia, and the Philippines). All times are given in local time. The Sun's altitude is given at the instant of maximum eclipse. *Eclipse Magnitude* is given as the fraction of the Sun's *diameter* covered by the Moon at maximum eclipse.

The 2028 Total Solar Eclipse Circumstances Calculator is an interactive web page that can quickly calculate the local circumstances for the solar eclipse from any geographic location not included in Tables 3 through 7. The URL address for the calculator is:

https://www.EclipseWise.com/solar/SEcirc/2001-2100/SE2028Jul22Tcirc.html

Table 3 – Total Eclipse Circumstances for Australia

Location	Partial Eclipse Begins	Total Eclipse Begins	Total Eclipse Ends	Partial Eclipse Ends	Duration Totality	Sun Altitude
Cocos Islands						
West Island	07:03:40 am	08:12:10 am	08:15:31 am	09:35:18 am	03m27s	24°
Christmas Island						
Fishing Fly Cove	07:39:45 am	08:54:52 am	08:58:46 am	10:25:41 am	03m53s	35°
Australia						
Durack, WA	09:27:27 am	10:58:08 am	11:03:12 am	12:32:47 pm	05m06s	53°
Kununurra, WA	09:27:53 am	10:59:45 am	11:02:38 am	12:33:14 pm	02m54s	53°
Purnululu N.P., WA	09:29:43 am	11:01:04 am	11:04:33 am	12:34:28 pm	03m28s	52°
Warmun, WA	09:28:22 am	10:59:13 am	11:03:41 am	12:33:26 pm	04m33s	52°
Wyndham, WA	09:25:54 am	10:57:27 am	11:00:42 am	12:31:31 pm	03m18s	53°
Karlu Karlu, NT	11:18:06 am	12:48:18 pm	12:53:13 pm	02:18:19 pm	04m55s	49°
Tennant Creek, NT	11:16:54 am	12:48:01 pm	12:51:25 pm	02:17:36 pm	03m30s	50°
Cordillo Downs, SA	11:41:59 am	01:09:37 pm	01:11:42 pm	02:32:03 pm	02m14s	41°
Bedourie, QLD	12:06:20 pm	01:34:09 pm	01:38:36 pm	02:59:20 pm	04m26s	44°
Bulloo Downs, QLD	12:19:18 pm	01:45:06 pm	01:47:24 pm	03:05:39 pm	02m27s	38°
Durham Downs, QLD	12:15:20 pm	01:41:15 pm	01:45:21 pm	03:03:52 pm	04m10s	40°
Eromanga, QLD	12:18:04 pm	01:44:39 pm	01:46:33 pm	03:05:25 pm	01m52s	40°
Eulo, QLD	12:23:39 pm	01:48:39 pm	01:50:52 pm	03:08:01 pm	02m13s	38°
Noccundra, QLD	12:17:46 pm	01:43:17 pm	01:47:00 pm	03:05:01 pm	03m48s	39°
Tanbar, QLD	12:13:59 pm	01:40:51 pm	01:44:00 pm	03:03:23 pm	03m14s	41°
Thargomindah, QLD	12:20:42 pm	01:45:24 pm	01:49:37 pm	03:06:33 pm	04m15s	38°
Barringun, NSW	12:25:55 pm	01:49:38 pm	01:52:59 pm	03:09:00 pm	03m24s	36°
Bathurst, NSW	12:37:24 pm	01:57:37 pm	02:00:39 pm	03:13:19 pm	03m03s	30°
Blacktown, NSW	12:10:08 pm	01:29:13 pm	01:32:58 pm	02:44:28 pm	03m49s	29°
Blayney, NSW	12:36:51 pm	01:58:02 pm	01:59:20 pm	03:13:00 pm	01m23s	30°
Bourke, NSW	12:27:22 pm	01:50:16 pm	01:54:17 pm	03:09:27 pm	04m06s	35°
Brewarrina, NSW	12:29:14 pm	01:52:01 pm	01:55:21 pm	03:10:25 pm	03m24s	35°
Byrock, NSW	12:28:52 pm	01:51:23 pm	01:55:10 pm	03:10:02 pm	03m51s	35°
Chatswood, NSW	12:10:41 pm	01:29:37 pm	01:33:22 pm	02:44:42 pm	03m48s	29°
Coolabah, NSW	12:29:51 pm	01:52:11 pm	01:55:44 pm	03:10:25 pm	03m39s	34°
Dubbo, NSW	12:34:39 pm	01:55:26 pm	01:59:13 pm	03:12:24 pm	03m50s	32°
Enngonia, NSW	12:26:29 pm	01:49:48 pm	01:53:36 pm	03:09:13 pm	03m47s	36°
Fords Bridge, NSW	12:25:59 pm	01:49:15 pm	01:53:16 pm	03:08:51 pm	04m05s	36°
Gilgandra, NSW	12:34:21 pm	01:55:20 pm	01:59:11 pm	03:12:25 pm	03m51s	32°
Girilambone, NSW	12:30:24 pm	01:52:38 pm	01:56:03 pm	03:10:38 pm	03m30s	34°
Gosford, NSW	12:40:47 pm	01:59:58 pm	02:03:11 pm	03:14:51 pm	03m16s	29°
Hornsby, NSW	12:40:31 pm	01:59:31 pm	02:03:16 pm	03:14:40 pm	03m47s	29°
Hungerford, NSW	12:23:03 pm	01:47:04 pm	01:51:07 pm	03:07:32 pm	04m06s	37°
Katoomba, NSW	12:38:57 pm	01:58:28 pm	02:01:55 pm	03:13:57 pm	03m32s	30°
Lithgow, NSW	12:38:32 pm	01:58:06 pm	02:01:44 pm	03:13:50 pm	03m41s	30°
Molong, NSW	12:35:48 pm	01:56:55 pm	01:59:06 pm	03:12:39 pm	02m18s	31°
Morisset, NSW	12:40:51 pm	02:00:53 pm	02:02:34 pm	03:14:58 pm	01m38s	29°
Mount Victoria, NSW	12:38:47 pm	01:58:18 pm	02:01:53 pm	03:13:55 pm	03m38s	30°
Mudgee, NSW	12:36:51 pm	01:57:01 pm	02:00:52 pm	03:13:19 pm	03m54s	31°
Nyngan, NSW	12:31:15 pm	01:53:19 pm	01:56:32 pm	03:10:57 pm	03m15s	33°
Orange, NSW	12:36:23 pm	01:57:20 pm	01:59:28 pm	03:12:52 pm	02m15s	31°
Parramatta, NSW	12:40:21 pm	01:59:22 pm	02:03:07 pm	03:14:33 pm	03m49s	29°
Penrith, NSW	12:39:12 pm	01:58:37 pm	02:02:07 pm	03:14:03 pm	03m34s	29°
Sydney, NSW	12:40:47 pm	01:59:41 pm	02:03:25 pm	03:14:44 pm	03m48s	29°
Wellington, NSW	12:35:34 pm	01:56:04 pm	01:59:48 pm	03:12:44 pm	03m47s	31°
Wollongong, NSW	12:40:31 pm	01:59:44 pm	02:02:40 pm	03:14:25 pm	02m55s	29°
Wyong, NSW	12:40:50 pm	02:00:19 pm	02:02:59 pm	03:14:55 pm	02m44s	29°
Yantabulla, NSW	12:24:40 pm	01:48:17 pm	01:52:22 pm	03:08:17 pm	04m09s	37°

Table 4 – Total Eclipse Circumstances for South Island, New Zealand

Location	Partial Eclipse Begins	Total Eclipse Begins	Total Eclipse Ends	Partial Eclipse Ends	Duration Totality	Sun Altitude
Alexandra	03:08:13 pm	04:15:35 pm	04:18:25 pm	05:20:13 pm	02m53s	9°
Balclutha	03:08:13 pm	04:15:41 pm	04:17:32 pm	05:19:39 pm	01m54s	9°
Dunedin	03:09:07 pm	04:15:56 pm	04:18:43 pm	-	02m51s	8°
Glenorchy	03:07:16 pm	04:15:07 pm	04:17:59 pm	05:20:11 pm	02m56s	10°
Kingston	03:07:27 pm	04:15:07 pm	04:17:47 pm	05:19:58 pm	02m42s	10°
Milton	03:08:29 pm	04:15:37 pm	04:18:02 pm	05:19:48 pm	02m29s	9°
Palmerston	03:09:29 pm	04:16:30 pm	04:18:45 pm	-	02m19s	8°
Queenstown	03:07:31 pm	04:15:13 pm	04:18:03 pm	05:20:09 pm	02m55s	10°
Wanaka	03:08:08 pm	04:15:53 pm	04:18:16 pm	05:20:30 pm	02m26s	10°

Table 5 - Partial Eclipse Circumstances for Australia

Location	Partial Eclipse Begins	Maximum Eclipse	Partial Eclipse Ends	Eclipse Magnitude	Sun Altitude
Adelaide, SA	11:47:54 am	01:11:26 pm	02:29:22 pm	0.803	34°
Albury, VIC	12:34:26 pm	01:55:45 pm	03:10:11 pm	0.909	29°
Alice Springs, NT	11:21:59 am	12:53:27 pm	02:19:54 pm	0.954	46°
Ballarat, VIC	12:30:23 pm	01:51:18 pm	03:06:03 pm	0.832	29°
Bendigo, VIC	12:30:21 pm	01:51:49 pm	03:06:53 pm	0.854	30°
Brisbane, QLD	12:41:10 pm	02:02:01 pm	03:14:37 pm	0.844	33°
Broken Hill, NSW	12:19:55 pm	01:45:22 pm	03:04:03 pm	0.907	36°
Broome, WA	09:16:16 am	10:46:33 am	12:19:53 pm	0.921	48°
Canberra, ACT	12:37:51 pm	01:58:53 pm	03:12:41 pm	0.961	29°
Carnarvon, WA	09:12:14 am	10:33:00 am	12:00:18 pm	0.708	36°
Charleville, QLD	12:24:48 pm	01:50:47 pm	03:08:35 pm	0.943	39°
Coffs Harbour, NSW	12:42:41 pm	02:03:16 pm	03:15:48 pm	0.910	31°
Coober Pedy, SA	11:31:51 am	01:00:04 pm	02:23:10 pm	0.858	41°
Darwin, NT	10:59:06 am	12:32:18 pm	02:03:18 pm	0.895	57°
Eucla, WA	09:54:11 am	11:19:02 am	12:41:27 pm	0.724	38°
Geelong, VIC	12:31:44 pm	01:52:09 pm	03:06:25 pm	0.829	29°
Hobart, TAS	12:40:09 pm	01:56:26 pm	03:07:12 pm	0.781	23°
Kalgoorlie, WA	09:38:05 am	11:00:09 am	12:23:31 pm	0.657	37°
Karratha, WA	09:10:11 am	10:35:40 am	12:07:10 pm	0.814	42°
Launceston, TAS	12:38:53 pm	01:56:20 pm	03:07:59 pm	0.806	24°
Mackay, QLD	12:28:18 pm	01:51:59 pm	03:07:11 pm	0.764	42°
Melbourne, VIC	12:32:30 pm	01:53:03 pm	03:07:20 pm	0.844	29°
Meekatharra, WA	09:24:13 am	10:47:36 am	12:14:38 pm	0.706	39°
Mildura, NSW	12:23:49 pm	01:47:33 pm	03:04:46 pm	0.872	33°
Moomba, SA	11:42:40 am	01:10:46 pm	02:31:52 pm	0.966	40°
Mount Isa, QLD	12:02:04 pm	01:33:03 pm	02:56:50 pm	0.938	48°
Newcastle, NSW	12:41:18 pm	02:02:06 pm	03:15:12 pm	0.994	29°
Parkes, NSW	12:34:29 pm	01:57:00 pm	03:11:59 pm	0.993	31°
Perth, WA	09:30:28 am	10:47:39 am	12:08:31 pm	0.591	33°
Rockhampton, QLD	12:33:05 pm	01:55:52 pm	03:10:12 pm	0.792	39°
Roxby Downs, SA	12:08:40 pm	01:35:37 pm	02:56:53 pm	0.856	38°
Southport, QLD	12:42:14 pm	02:02:47 pm	03:15:07 pm	0.849	33°
Toowoomba, QLD	12:38:46 pm	02:00:30 pm	03:13:57 pm	0.866	34°
Townsville, QLD	12:20:32 pm	01:46:06 pm	03:03:19 pm	0.765	45°
Yulara, NT	11:17:38 am	12:48:04 pm	02:14:56 pm	0.877	45°
Wagga Wagga, NSW	12:34:29 pm	01:56:15 pm	03:10:54 pm	0.937	30°
Whyalla, SA	11:43:30 am	01:08:37 pm	02:28:07 pm	0.824	36°

Table 6 - Partial Eclipse Circumstances for New Zealand

Location	Partial Eclipse Begins	Maximum Eclipse	Partial Eclipse Ends	Eclipse Magnitude	Sun Altitude
Auckland	03:18:27 pm	04:23:39 pm	05:23:07 pm	0.737	10°
Christchurch	03:12:18 pm	04:19:52 pm	-	0.938	8°
Hamilton	03:18:31 pm	04:23:41 pm	-	0.754	10°
Invercargill	03:06:46 pm	04:15:39 pm	05:19:07 pm	0.985	9°
Lower Hutt	03:15:59 pm	04:22:13 pm	-	0.850	8°
Manukau	03:18:31 pm	04:23:41 pm	05:23:08 pm	0.739	10°
Napier	03:19:18 pm	04:23:56 pm	-	0.779	8°
Rotorua	03:19:26 pm	04:24:05 pm	-	0.752	9°
Takapuna	03:18:32 pm	04:23:41 pm	05:23:06 pm	0.734	10°
Tauranga	03:19:39 pm	04:24:11 pm	-	0.740	9°
Waitemata	03:18:20 pm	04:23:36 pm	05:23:08 pm	0.739	10°
Wellington	03:15:47 pm	04:22:07 pm	-	0.854	8°

Table 7 - Partial Eclipse Circumstances for Asia, Indonesia and Philippines

Country	City	Partial Eclipse Begins	Maximum Eclipse	Partial Eclipse Ends	Eclipse Magnitude	Sun Altitude
Cambodia	Phnom Penh	07:45:41 am	08:41:28 am	09:43:04 am	0.353	40°
India	Bangalore	06:24:33 am	06:50:24 am	07:17:21 am	0.103	10°
	Madras	06:22:03 am	06:51:08 am	07:21:41 am	0.127	13°
Indonesia	Bandung	08:39:22 am	09:57:10 am	11:26:30 am	0.922	38°
	Jakarta	08:38:05 am	09:54:55 am	11:23:18 am	0.901	38°
	Surabaya	08:46:18 am	10:08:39 am	11:41:17 am	0.928	45°
Malaysia	Kuala Lumpur	07:34:04 am	08:39:58 am	09:55:12 am	0.607	34°
Myanmar	Yangon	07:26:19 am	08:00:57 am	08:37:42 am	0.140	31°
New Caledonia	Noumea	02:12:47 pm	03:15:45 pm	04:12:47 pm	0.468	26°
Papua New Guinea	Port Moresby	12:17:06 pm	01:35:41 pm	02:47:23 pm	0.516	55°
Philippines	Manila	09:13:40 am	10:07:28 am	11:03:57 am	0.236	62°
	Quezon City	09:13:54 am	10:07:31 am	11:03:48 am	0.234	62°
Singapore	Singapore	08:34:41 am	09:44:11 am	11:03:48 am	0.674	36°
Sri Lanka	Colombo	06:08:04 am	06:50:59 am	07:37:43 am	0.316	10°
Thailand	Bangkok	07:47:58 am	08:35:28 am	09:27:15 am	0.264	35°

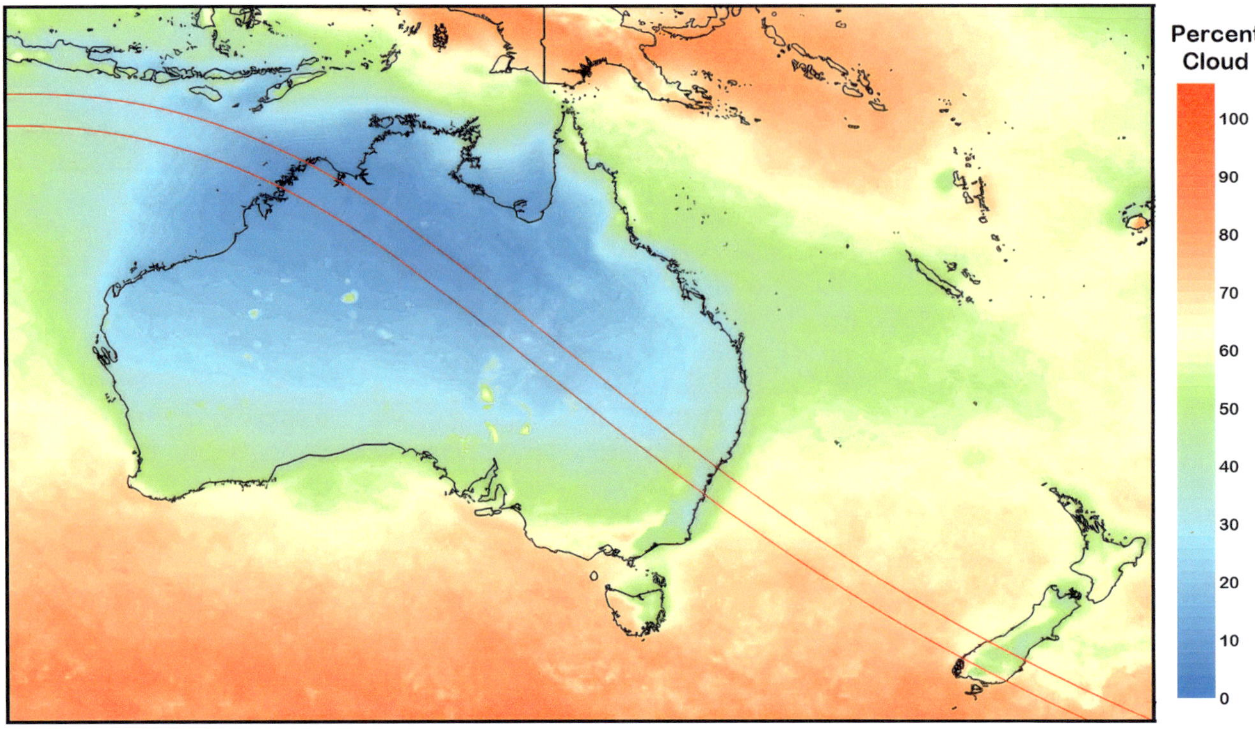

Figure 10 – July 22, 2028 total solar eclipse with average July daytime cloud cover derived from 20 years of satellite data, 2000-2019. Data: CM-SAF. (map by Jay Anderson)

6 – Weather Prospects for the 2028 Total Solar Eclipse

The 2028 total eclipse track crosses Australia and New Zealand. Northern Australia offers excellent prospects for clear sky during the eclipse. Southeastern Australia (including Sydney) and New Zealand have a higher probability of clouds, but these regions also have better infrastructure and are easier to reach.

Table 8 lists sixteen locations in the path of totality and include the duration of totality, the average high temperature, and the average cloud cover during July.

Table 8 – Average High Temperature and Cloud Cover in July

Country	Location	Duration Totality	Ave. High Temp.[1]	Ave. Cloud Cover[2]
Australia	Cocos Islands	3m27s	28°C (82°F)	65%
Australia	Christmas Island	3m 53s	27°C (81°F)	57%
Australia	Mitchell River Nat. Park	5m 10s	38°C (100°F)	8%
Australia	Kalkarindji	4m 10s	28°C (83°F)	16%
Australia	Tennant Creek	3m 30s	26°C (78°F)	20%
Australia	Bedourie	4m 26s	21°C (69°F)	17%
Australia	Durham Downs	4m 10s	19°C (66°F)	37%
Australia	Thargomindah	4m 15s	20°C (68°F)	24%
Australia	Bourke	4m 6s	18°C (65°F)	44%
Australia	Dubbo	3m 50s	18°C (65°F)	43%
Australia	Wellington	3m 47s	16°C (60°F)	44%
Australia	Bathurst	3m 3s	12°C (54°F)	57%
Australia	Katoomba	3m 32s	10°C (50°F)	53%
Australia	Sydney	3m 48s	17°C (63°F)	47%
New Zealand	Queenstown	2m 55s	8°C (46°F)	62%
New Zealand	Dunedin	2m 51s	9°C (49°F)	59%

[1]*Average High Temperature in July [source: www.weatherspark.com]*

[2]*Average Cloud Cover in July (since 2000) [source: www.timeanddate.com]*

7 – Saros 146

The total eclipse of 2028 is part of a series of eclipses producing relatively long totalities during the 20th and 21st centuries. These exceptional eclipses all have durations exceeding four minutes and belong to a group called Saros 146.

The Saros is a period in which eclipses repeat with very similar characteristics. It has a duration of approximately 6,585.3 days (~18 years 11 days 8 hours). When two eclipses are separated by a period of one Saros, they share a very similar geometry. The eclipses occur at the same node with the Moon at nearly the same distance from Earth and at the same time of year. Thus, the Saros is useful for organizing eclipses into families or series. A Saros series doesn't last forever because the characteristics of the eclipse geometry slowly change with each cycle. A series typically lasts 12 or 13 centuries and contains 70 or more eclipses.

The total eclipse of 2028 is the 28th member of Saros series 146. The Saros series were first numbered by van den Bergh in 1955. All eclipses in an even numbered Saros series occur at the Moon's descending node and the Moon moves northward with each succeeding member in the family (i.e., gamma[6] increases). Saros 146 is a relatively young series. The series began with a small partial eclipse visible off the coast of Antarctica on Sept. 19, 1541.

After an unusually long sequence of 21 more partial eclipses, the first central eclipse occurred on May 29, 1938. It was total with a maximum duration of 4.1 minutes. Subsequent total eclipses in the series have seen an increase in the duration of totality. The total eclipse of July 11, 2010 marked the longest totality of Saros 146 because the durations of all future events will decrease. Nevertheless, the 2028 eclipse still produces a duration 5m 10s which is far exceeds the average duration of about 2.5 minutes.

The series produces the first of 4 hybrid eclipses on Oct 17, 2172. The remaining 24 central eclipses of Saros 146 are all annular and span the period from Dec 01, 2244 to Aug 10, 2659. The series ends with a set of 13 partial eclipses the last of which occurrs on Dec 29, 2893.

In summary, Saros series 146 includes 76 eclipses. It begins with 22 partial eclipses, followed by 13 total, 4 hybrid, and 24 annular eclipses. It ends with a string of 13 partial eclipses. From start to finish, the series spans a period of 1352 years. The EclipseWise.com catalog for Saros 146 has more details and maps for this series:

www.eclipsewise.com/solar/SEsaros/SEsaros146.html

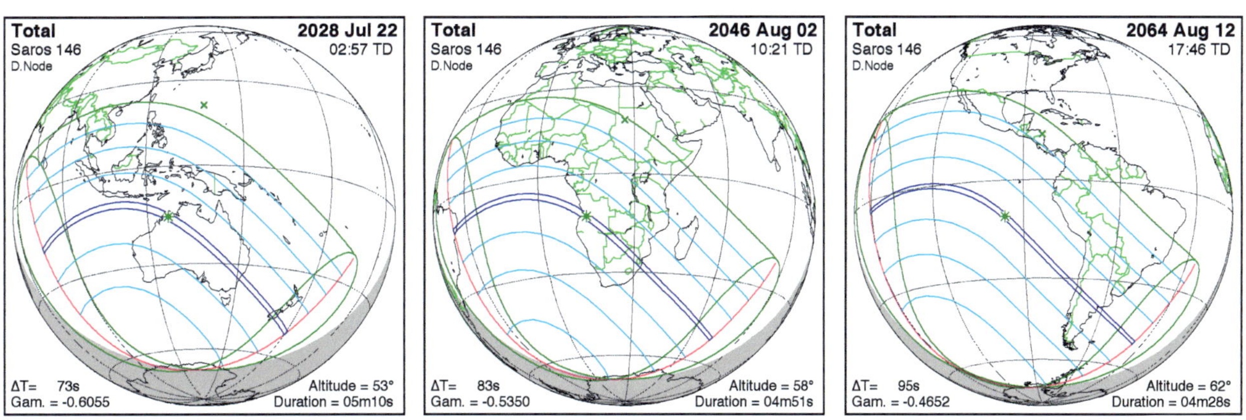

Figure 11 – Three total solar eclipses of Saros 146 for the years 2028 (5m 10s), 2046 (4m 51s), and 2064 (4m 26s).

---

6 *Gamma is the distance of the Moon's shadow axis from Earth's center in units of equatorial Earth radii. It is defined at the instant of greatest eclipse when its absolute value is at a minimum.*

8 – Detailed Maps of the Path of Totality

A series of 19 full-page maps (pages 25-43) cover the entire 2028 path of totality through Australia and New Zealand. The *Table of Contents* (page iii) can be used to quickly navigate to the map of interest since it lists the region featured on each map. North is up on each map.

The map scale is approximately 1:1,400,000, which corresponds to 1 inch ≈ 22 miles (1 cm ≈ 14 km).[7] This large scale shows both major and minor roads, towns and cities, rivers, lakes, parks, national forests, wilderness areas and mountain ranges. Two reference scales for 50 miles and 50 kilometers appear at the bottom of each map.

Two maps focus on Sydney and use a larger scale of approximately 1:700,000 (1 inch ≈ 11 miles; 1 cm ≈ 7 km) and 1:350,000 (1 inch ≈ 5.5 miles; 1 cm ≈ 3.5 km).

The path of totality on each map is depicted as a lightly shaded region with the northern and southern limits clearly labeled. The total eclipse can be seen only inside this path (a partial eclipse is visible outside the path). The closer one gets to the central line, the longer the total eclipse lasts. Gray lines inside the path mark the duration of the total eclipse in 30-second steps. This makes it easy to estimate the duration of totality from any location in the eclipse path.

The local time of mid-eclipse is marked by a series of white lines crossing the eclipse path every five minutes. Abbreviations for local times are: CCT, CXT, AWST, ACST, AEST, and NZST (see Table 1). Eclipse circumstances on the central line are labeled with the local time of mid-eclipse, duration of totality (minutes and seconds) and altitude of the Sun.

All maps were produced using Google Maps as the underlying map with overlying eclipse graphics generated using JavaScript code. A web page is available to the user for examining any part of the 2028 eclipse path at a range of zoom magnifications. An added benefit of the web page is that it automatically calculates the local circumstances for any point chosen by the user, with just the click of a mouse. For more information and to access the interactive 2028 eclipse path plotted on Google Maps, visit:

https://www.eclipsewise.com/solar/SEgmapx/2001-2100/SE2028Jul22Tgmapx.html#map

9 – EclipseWise.com Web Site

EclipseWise.com has individual web pages, maps and diagrams for every solar and lunar eclipse covering the 5000-year period from 2000 BCE to 3000 CE. This includes 11,898 solar eclipses and 12,064 lunar eclipses. Much of the design, layout and graphics were inspired by the recent publications *Thousand Year Canon of Solar Eclipses 1501 to 2500* and the *Thousand Year Canon of Lunar Eclipses 1501 to 2500.* For more information on these books, see: ***https://www.astropixels.com/pubs***

The graphical user interface used by *EclipseWise.com* offers an intuitive way of accessing eclipse predictions. For example, the home page presents a concise preview of all upcoming solar and lunar eclipses over several years. Each small eclipse diagram gives a quick preview of an eclipse and links to a dedicated page for that eclipse.

The main or top pages of EclipseWise.com are:

Home Page (both solar and lunar eclipses): ***https://www. eclipsewise.com/eclipse.html***
Solar Eclipses Page: ***https://www. eclipsewise.com/solar/solar.html***
Lunar Eclipses Page: ***https://www. eclipsewise.com/lunar/lunar.html***

[7] Because of the Mercator map projection, the actual scale on a given map can vary by up to 10% of this value.

10 – EclipseWise.com and the 2028 Eclipse

EclipseWise.com has a series of pages and resources devoted to the 2028 eclipse. The main page is located at:

https://www.eclipsewise.com/solar/SEprime/2001-2100/SE2028Jul22Tprime.html

It provides links to detailed eclipse path maps, tables of eclipse path coordinates, Besselian elements, and more. The link to an interactive Google Map with the eclipse path plotted on it allows the user to zoom into a part of the path. Click on any point on the map to display the eclipse circumstances and duration of totality at that location.

Other features include information on eye safety, eclipse photography, the sky during totality, and additional data tables about the eclipse path. This web site will continue to add features as the eclipse approaches.

The *2028 Total Solar Eclipse Circumstances Calculator* is an interactive web page that can quickly calculate the local circumstances for the eclipse from any geographic location. The *Calculator* is located at:

https://www.eclipsewise.com/solar/SEcirc/2001-2100/SE2028Jul22Tcirc.html

11 – Eclipse Predictions

The algorithms and software for the eclipse predictions were developed primarily from the *Explanatory Supplement to the Astronomical Ephemeris* (Her Majesty's Nautical Almanac Office, 1974) with additional algorithms from *Elements of Solar Eclipses: 1951–2200* (Meeus, 1989). The solar and lunar ephemerides were generated from the JPL DE430. All eclipse calculations were made using a value for the Moon's radius of k=0.2722810 for the path of totality. Eclipse contact times for locations and cities (Tables 3 and 4) have been corrected for the effects of the Moon's limb profile. A value for ΔT of 69.2 seconds was used to convert the predictions from Terrestrial Time (TT) to Coordinated Universal Time (UTC).

Photo 2 – Viewing the Total Solar Eclipse of 2019 July 03.
Vicuna, Chile, © 2019 F. Espenak, www.MrEclipse.com

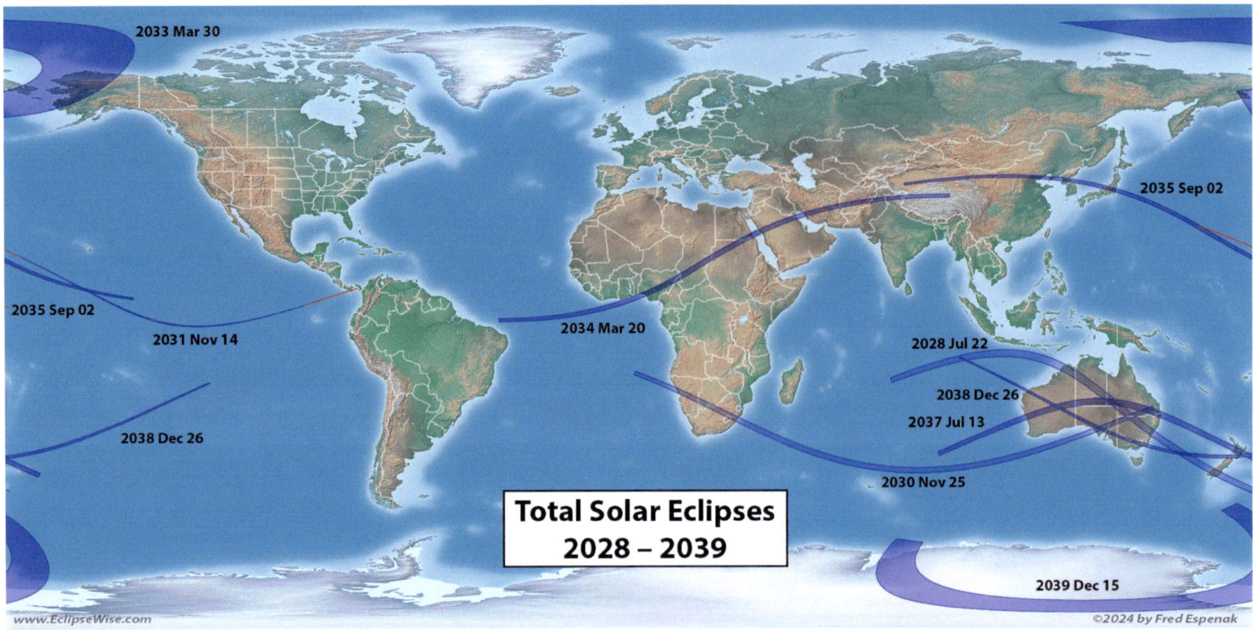

Figure 12 – The path of every total solar eclipse from 2028 through 2039 is plotted on a world map.

12 – Total Solar Eclipses: 2028 – 2039

During the 12-year period 2028 through 2039, Earth will experience nine total solar eclipses. Examination of the map in Figure 12 shows that only one of them is visible from North America (Alaska on 2033 Mar 30). In contrast, Australia is fortunate to have no less than four total solar eclipses during the same period (2028 July 22, 2030, Nov 25, 2037 July 13, and 2038 Dec 26). Information for each eclipse appears in *Table 9*.

Table 9 – Data for Total Solar Eclipses

Calendar Date	Time of Greatest Eclipse	Saros Number	Latitude	Longitude	Sun Alt	Path Width km	Duration Totality
2028 Jul 22	02:55:29	146	16°S	127°E	53°	230	05m10s
2030 Nov 25	06:50:27	133	44°S	71°E	67°	169	03m44s
2031 Nov 14*	21:06:20	143	1°S	138°W	72°	38	01m08s
2033 Mar 30	18:01:25	120	71°N	156°W	11°	781	02m37s
2034 Mar 20	10:17:35	130	16°N	22°E	73°	159	04m09s
2035 Sep 02	01:55:36	145	29°N	158°E	68°	116	02m54s
2037 Jul 13	02:39:25	127	25°S	139°E	43°	201	03m58s
2038 Dec 26	00:58:58	142	40°S	164°E	73°	95	02m18s
2039 Dec 15	16:22:34	152	81°S	173°E	18°	380	01m51s

* The eclipse of 2031 Nov 14 is a Hybrid Eclipse or Annular/Total Eclipse.

Explanation of Column Headings in Table 9

Calendar Date – The UTC date of the eclipse at the instant of Greatest Eclipse
Time of Greatest Eclipse – The UTC Time of Greatest Eclipse
Saros Number – The number of the Saros series that the eclipse belongs to
Latitude – The latitude of the Moon's umbral shadow at instant of Greatest Eclipse
Longitude – The longitude of the Moon's umbral shadow at instant of Greatest Eclipse
Sun Alt – The altitude of the Sun at the instant of Greatest Eclipse
Path Width – The width of the path of totality (km) at the instant of Greatest Eclipse
Duration Totality – The central line duration of totality at the instant of Greatest Eclipse

Total Solar Eclipses from 2028 to 2039

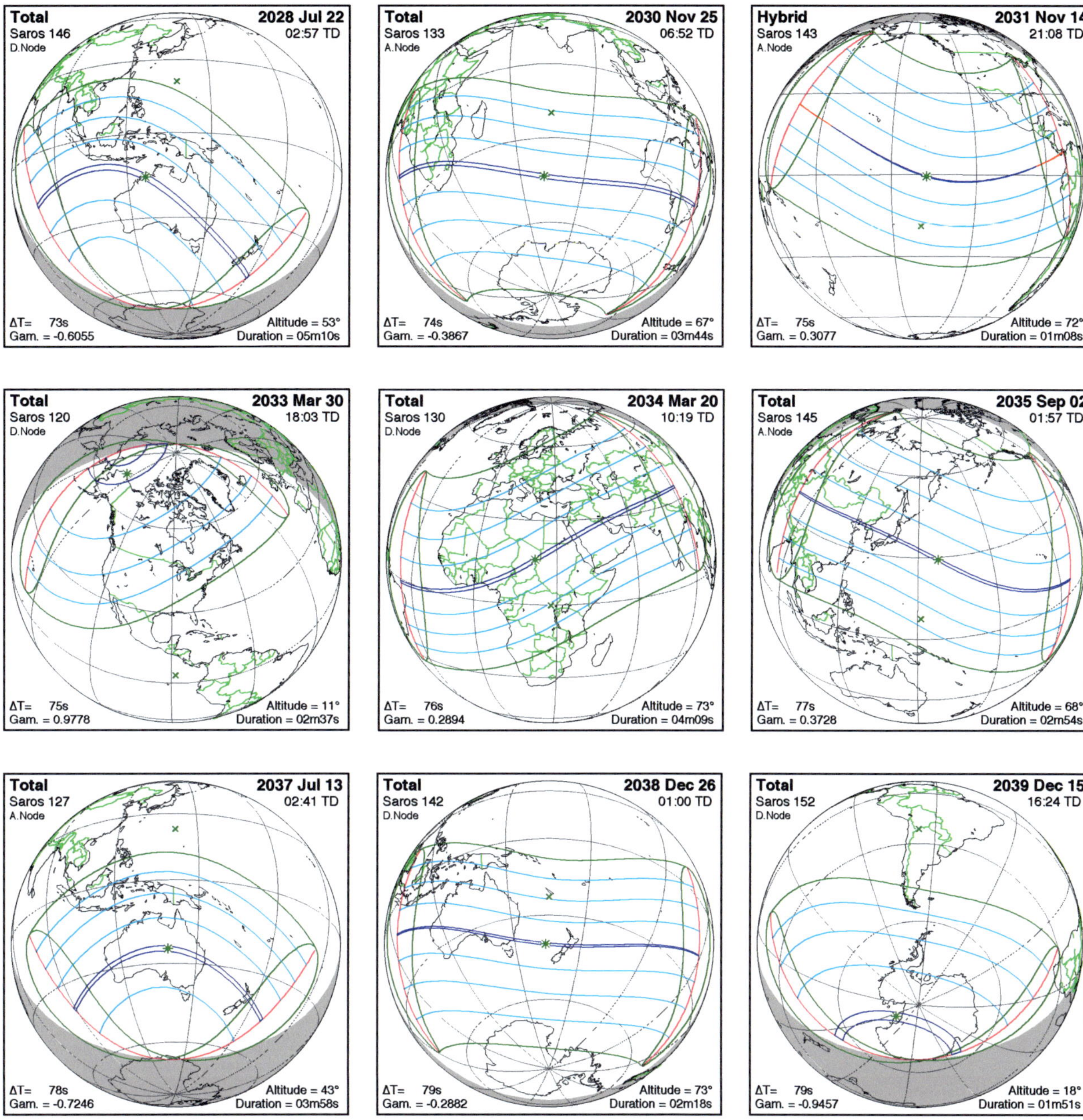

Figure 13 – Total Solar Eclipses from 2028 through 2039.
*Courtesy of **Thousand Year Canon of Solar Eclipses1501 to 2500**, F. Espenak, (2014)*

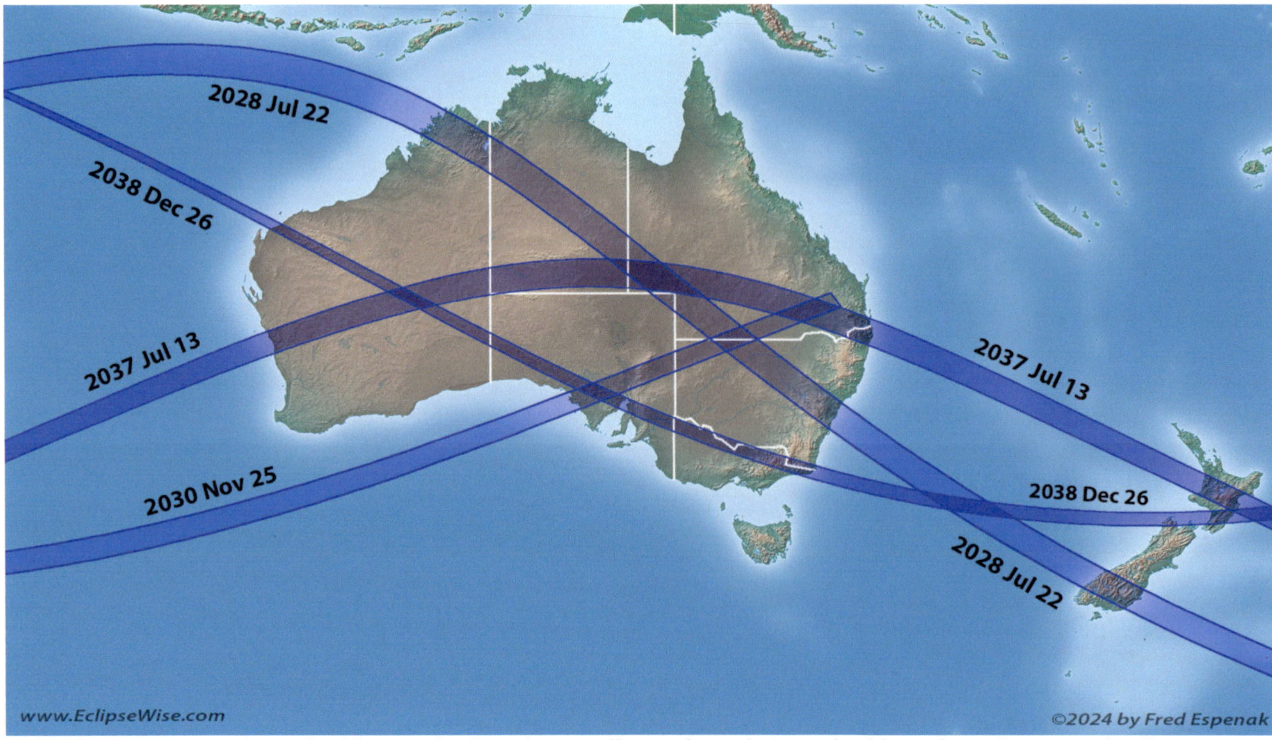

Figure 14 – The path of four total solar eclipses will pass through Australia (three through New Zealand) between 2028 and 2038.

13 – Total Solar Eclipses in Australia and New Zealand: 2028 – 2039

During the 11-year period 2028 through 2038, the path of four total solar eclipses will pass through Australia. The path of three of these eclipses also cross New Zealand.

1) July 22, 2028† - Path crosses the Kimberley in Western Australia, Northern Territory, southwest Queensland, New South Wales and passing centrally through Sydney. Maximum Duration in Australia: 5m 10s

2) November 25, 2030 - Path crosses South Australia, northwest New South Wales, southern Queensland ending at sunset in southeast Queensland. Maximum Duration in Australia: 2m 1s

3) July 13, 2037† - Path crosses southern Western Australia, southern Northern Territory, western Queensland and finally passing over Brisbane and the Gold Coast. Maximum Duration in Australia: 3m 59s

4) December 26, 2038† - Path crosses central Western Australia, South Australia, and along the New South Wales/Victoria border. Maximum Duration in Australia: 2m 10s

---

† Path of totality also crosses New Zealand

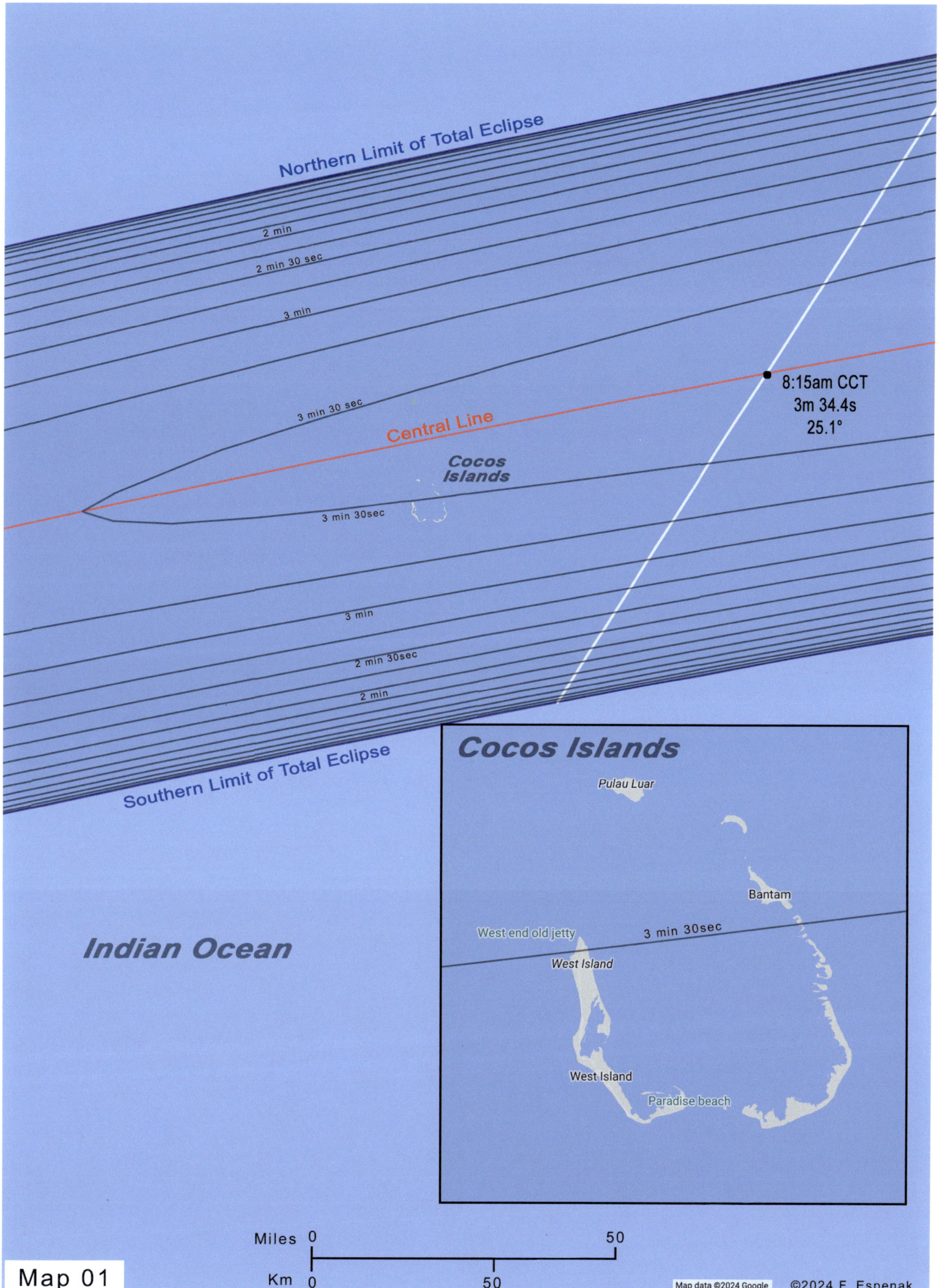
Northern Limit of Total Eclipse
2 min
2 min 30 sec
3 min
3 min 30 sec
Central Line
8:15am CCT
3m 34.4s
25.1°
Cocos Islands
3 min 30sec
3 min
2 min 30sec
2 min
Southern Limit of Total Eclipse
Cocos Islands
Pulau Luar
Bantam
West end old jetty
3 min 30sec
West Island
West Island
Paradise beach
Indian Ocean
Miles 0 50
Km 0 50
Map 01
Map data ©2024 Google
©2024 F. Espenak

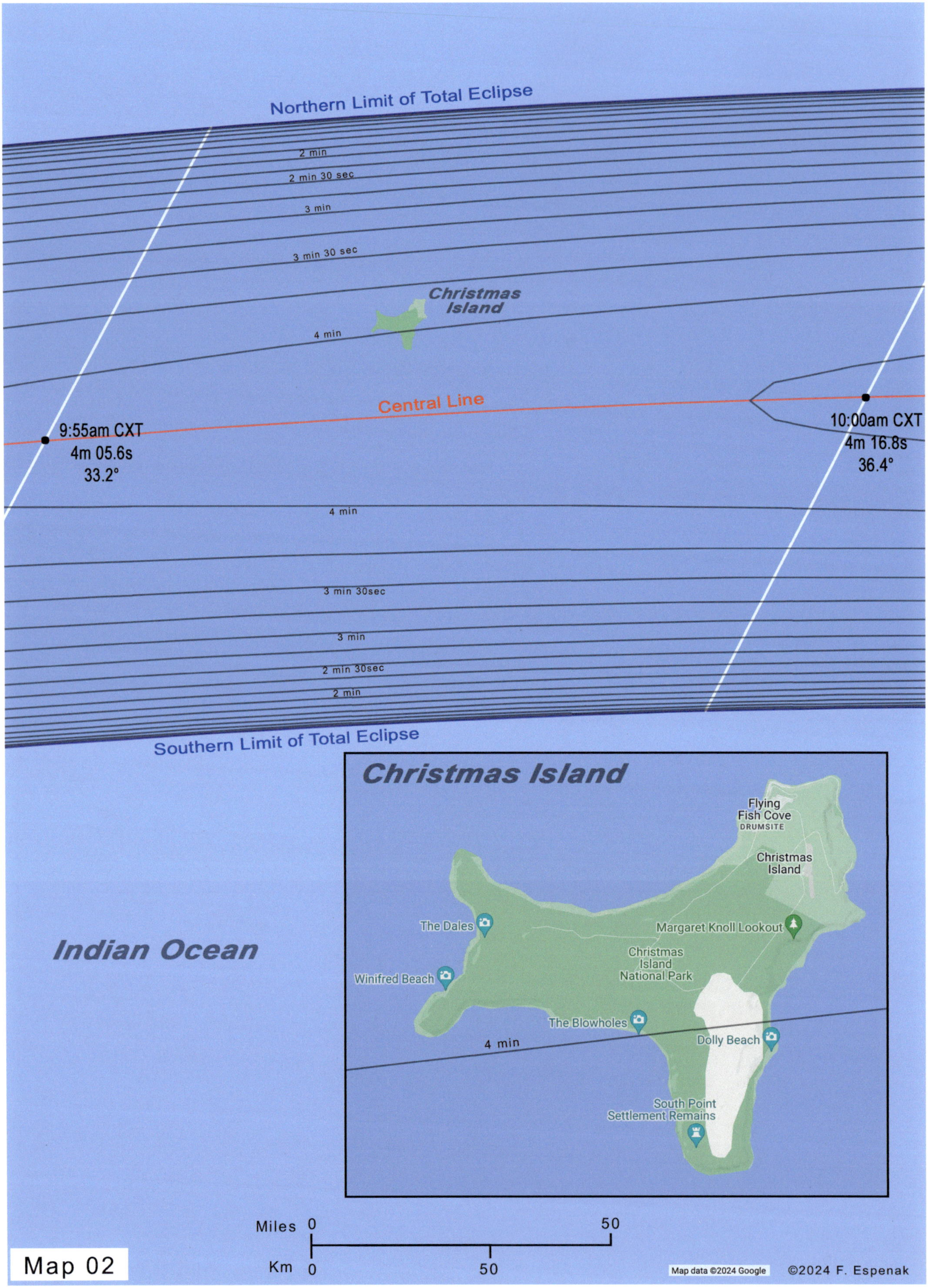
Northern Limit of Total Eclipse
2 min
2 min 30 sec
3 min
3 min 30 sec
Christmas Island
4 min
Central Line
9:55am CXT
4m 05.6s
33.2°
10:00am CXT
4m 16.8s
36.4°
4 min
3 min 30sec
3 min
2 min 30sec
2 min
Southern Limit of Total Eclipse
Indian Ocean
Christmas Island
Flying Fish Cove
DRUMSITE
Christmas Island
The Dales
Margaret Knoll Lookout
Christmas Island National Park
Winifred Beach
The Blowholes
Dolly Beach
4 min
South Point Settlement Remains
Miles 0 50
Km 0 50
Map 02
Map data ©2024 Google
©2024 F. Espenak

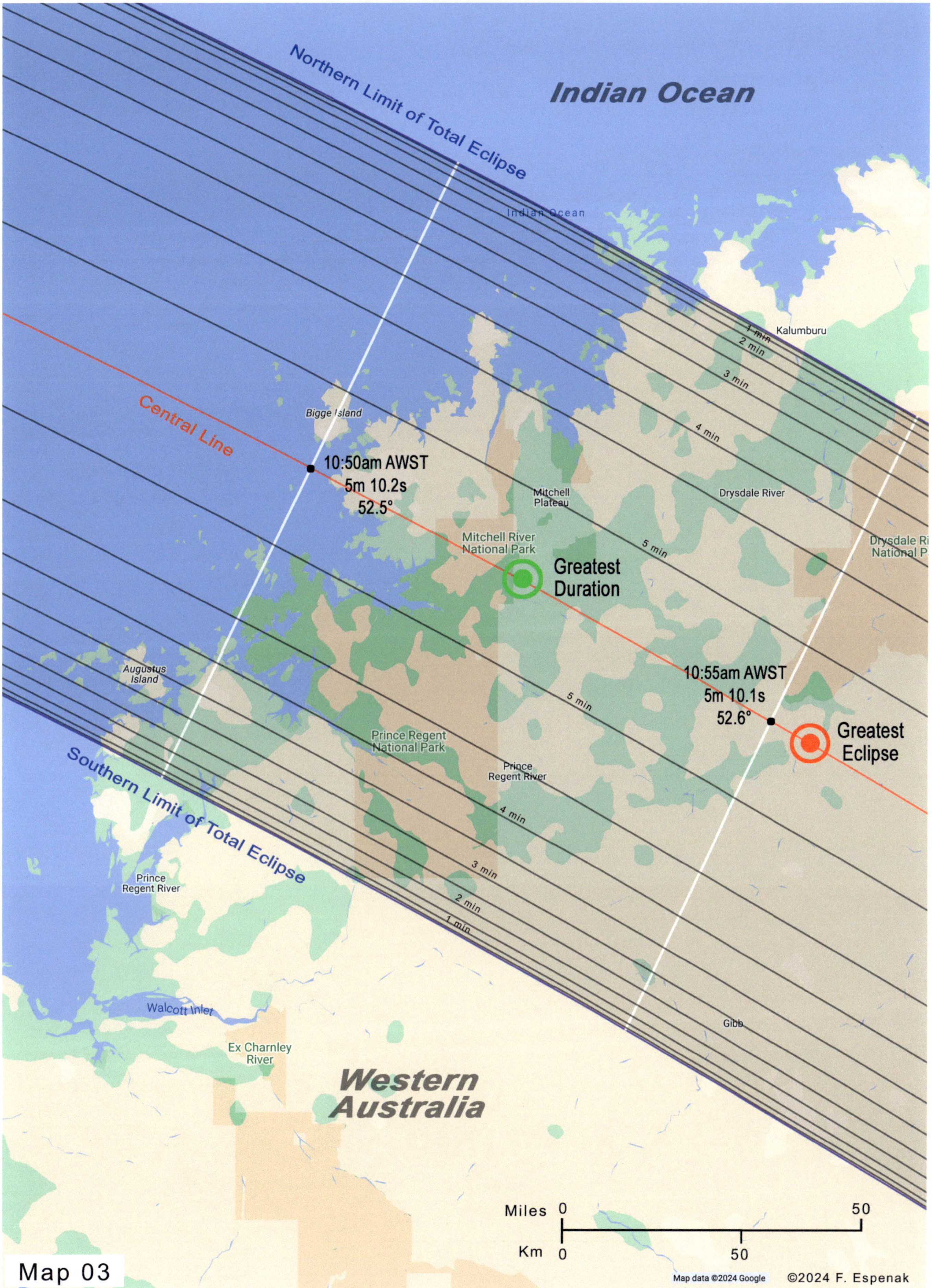
Indian Ocean
Northern Limit of Total Eclipse
Indian Ocean
Kalumburu
1 min
2 min
3 min
4 min
5 min
Central Line
Bigge Island
10:50am AWST
5m 10.2s
52.5°
Mitchell Plateau
Drysdale River
Mitchell River National Park
Greatest Duration
Augustus Island
10:55am AWST
5m 10.1s
52.6°
Greatest Eclipse
5 min
Prince Regent National Park
Prince Regent River
4 min
3 min
2 min
1 min
Southern Limit of Total Eclipse
Prince Regent River
Walcott Inlet
Gibb
Ex Charnley River
Western Australia
Miles 0 50
Km 0 50
Map 03
Map data ©2024 Google
©2024 F. Espenak

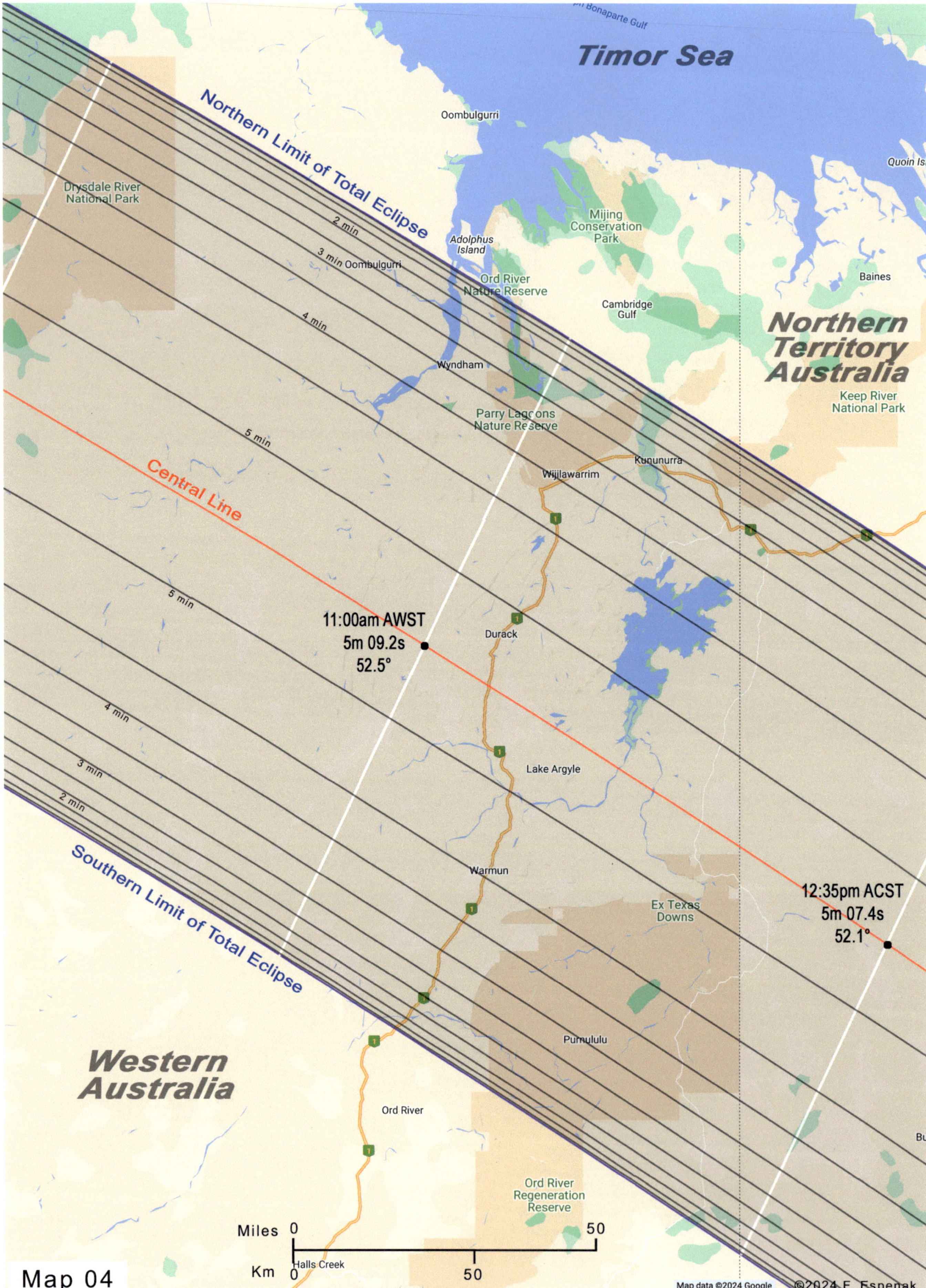
Timor Sea
Bonaparte Gulf
Northern Limit of Total Eclipse
Oombulgurri
Quoin Isla
Drysdale River National Park
2 min
3 min Oombulgurri
Adolphus Island
Mijing Conservation Park
Ord River Nature Reserve
Cambridge Gulf
Baines
4 min
Northern Territory Australia
Wyndham
Keep River National Park
Parry Lagoons Nature Reserve
5 min
Central Line
Kununurra
Wijilawarrim
5 min
11:00am AWST
5m 09.2s
52.5°
Durack
4 min
Lake Argyle
3 min
2 min
Southern Limit of Total Eclipse
Warmun
Ex Texas Downs
12:35pm ACST
5m 07.4s
52.1°
Purnululu
Western Australia
Ord River
Buc
Ord River Regeneration Reserve
Miles 0 50
Km 0 50
Halls Creek
Map 04
Map data ©2024 Google
©2024 F. Espenak

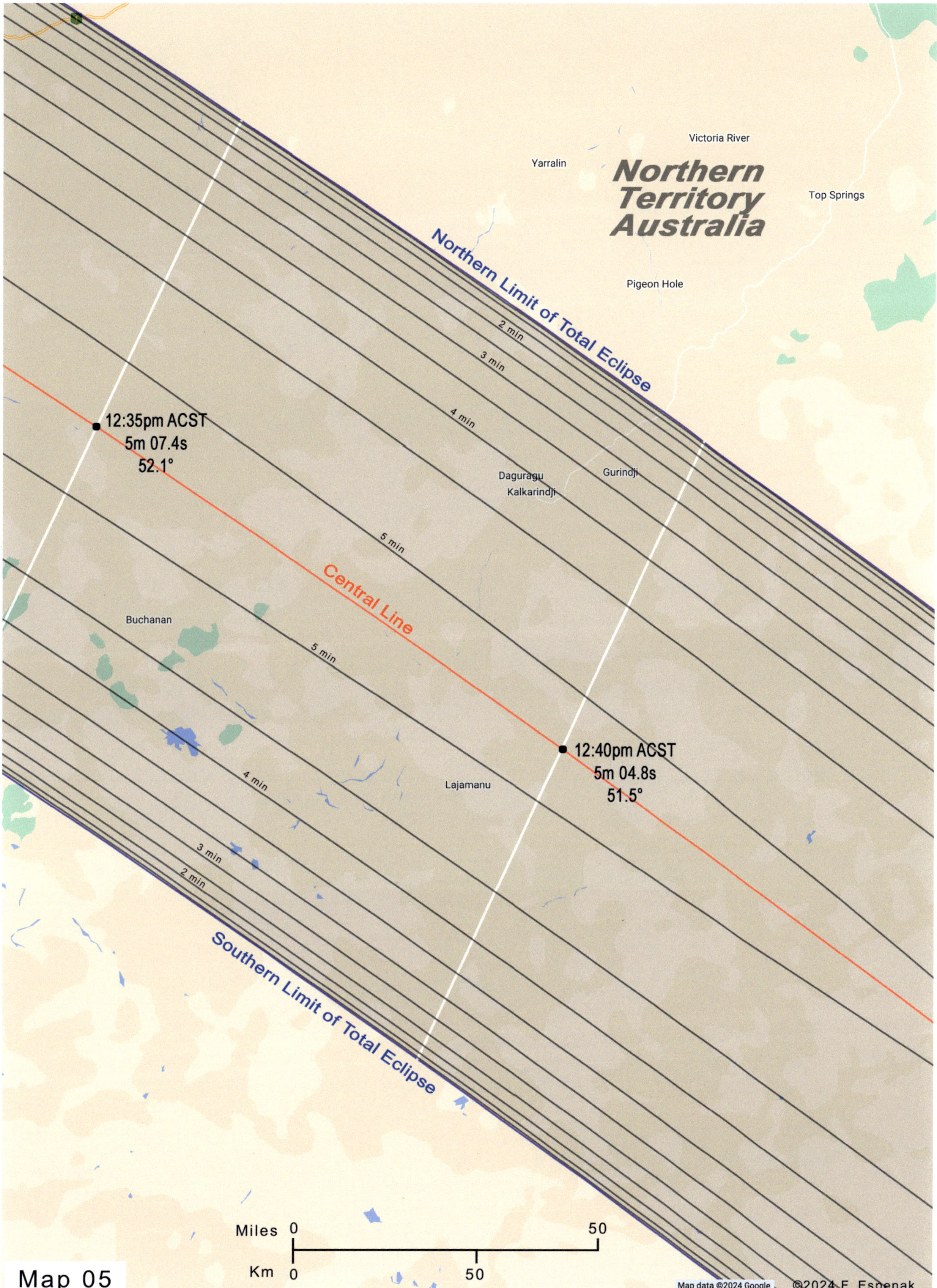
Victoria River
Yarralin
Northern Territory Australia
Top Springs
Northern Limit of Total Eclipse
Pigeon Hole
2 min
3 min
4 min
12:35pm ACST
5m 07.4s
52.1°
Daguragu
Kalkarindji
Gurindji
5 min
Central Line
Buchanan
5 min
12:40pm ACST
5m 04.8s
51.5°
4 min
Lajamanu
3 min
2 min
Southern Limit of Total Eclipse
Miles 0 50
Km 0 50
Map 05
Map data ©2024 Google
©2024 F. Espenak

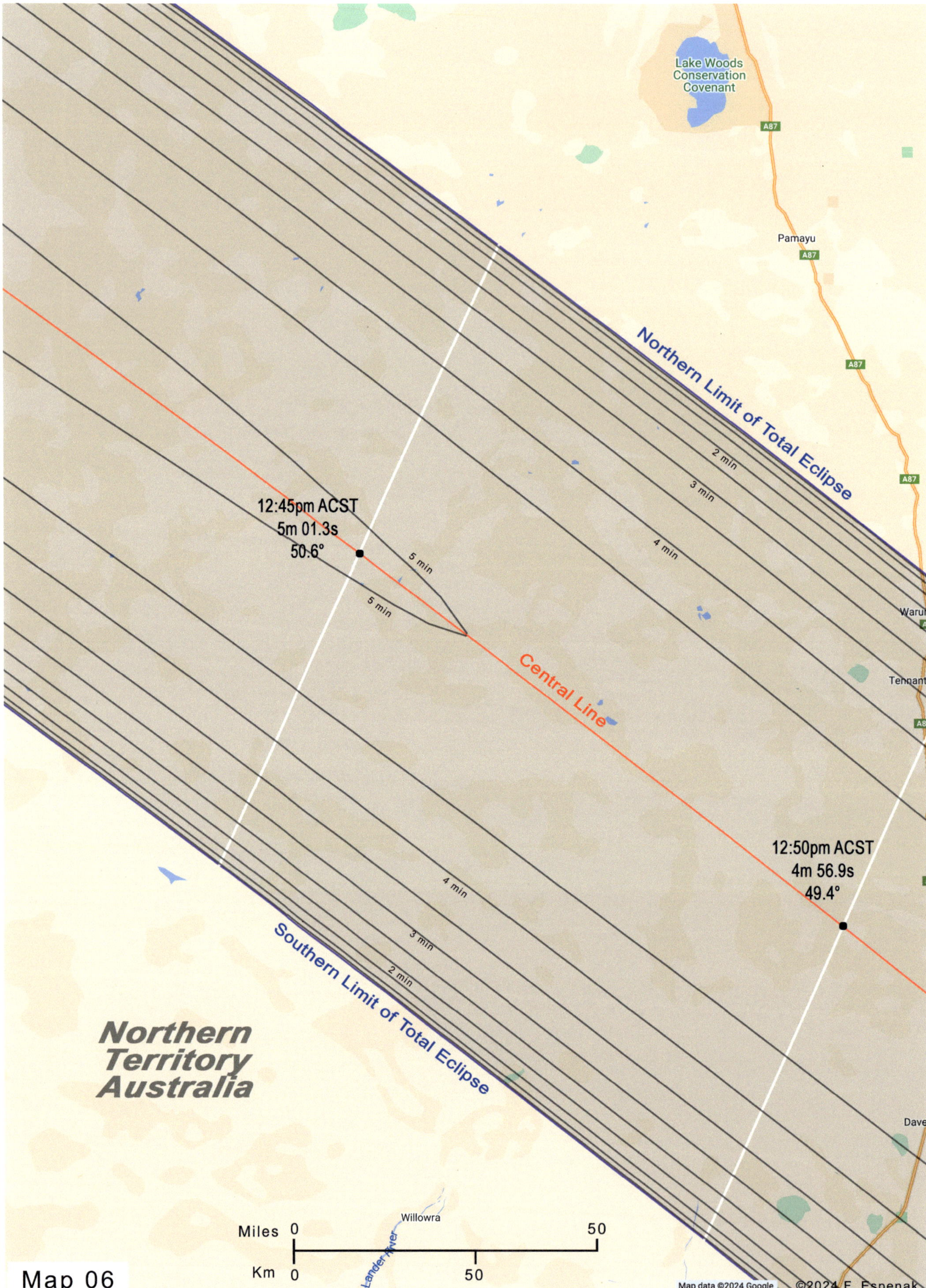
Lake Woods Conservation Covenant
A87
Pamayu
Northern Limit of Total Eclipse
2 min
3 min
4 min
5 min
5 min
12:45pm ACST
5m 01.3s
50.6°
Central Line
12:50pm ACST
4m 56.9s
49.4°
4 min
3 min
2 min
Southern Limit of Total Eclipse
Northern Territory Australia
Willowra
Lander River
Miles 0 50
Km 0 50
Map 06
Map data ©2024 Google
©2024 F. Espenak

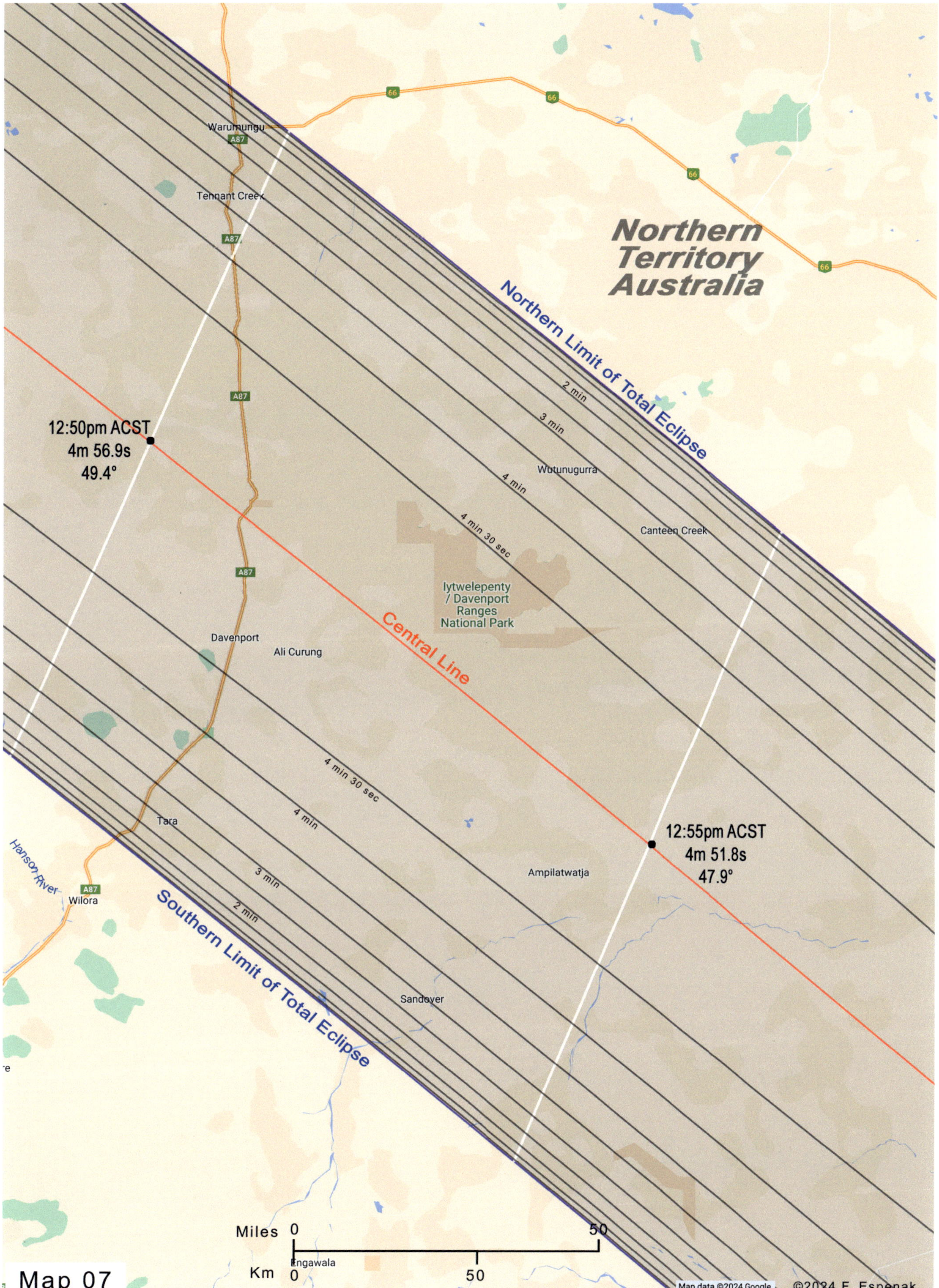
Northern Territory Australia
Northern Limit of Total Eclipse
Southern Limit of Total Eclipse
Central Line
12:50pm ACST
4m 56.9s
49.4°
12:55pm ACST
4m 51.8s
47.9°
2 min
3 min
4 min
4 min 30 sec
Warumungu
Tennant Creek
Wutunugurra
Canteen Creek
Iytwelepenty / Davenport Ranges National Park
Davenport
Ali Curung
Tara
Wilora
Hanson River
Ampilatwatja
Sandover
Engawala
Miles 0 50
Km 0 50
Map 07
Map data ©2024 Google
©2024 F. Espenak

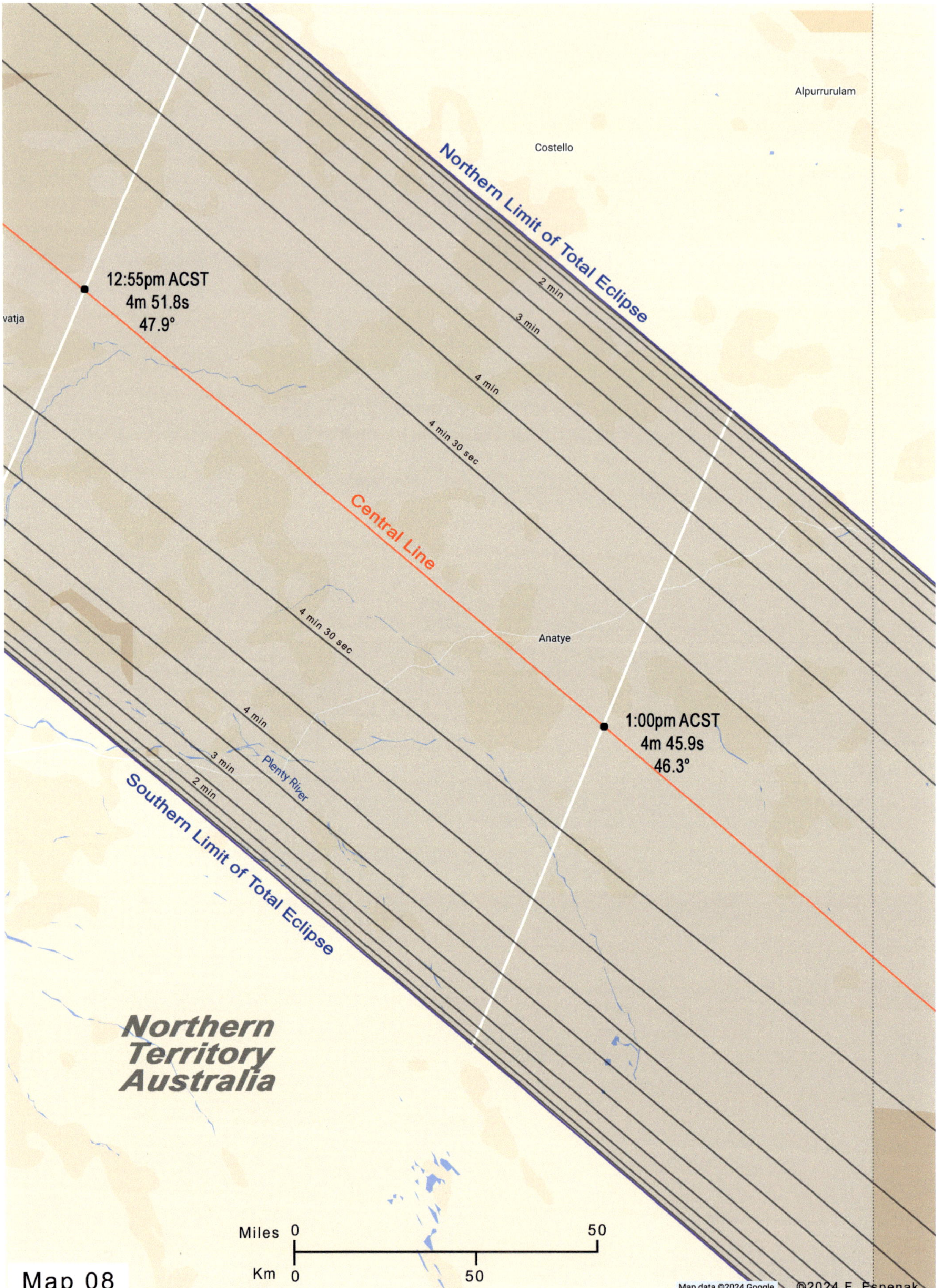
Alpurrurulam
Costello
Northern Limit of Total Eclipse
2 min
3 min
4 min
4 min 30 sec
12:55pm ACST
4m 51.8s
47.9°
vatja
Central Line
Anatye
4 min 30 sec
4 min
3 min
2 min
Plenty River
1:00pm ACST
4m 45.9s
46.3°
Southern Limit of Total Eclipse
Northern Territory Australia
Miles 0 50
Km 0 50
Map 08
Map data ©2024 Google
©2024 F. Espenak

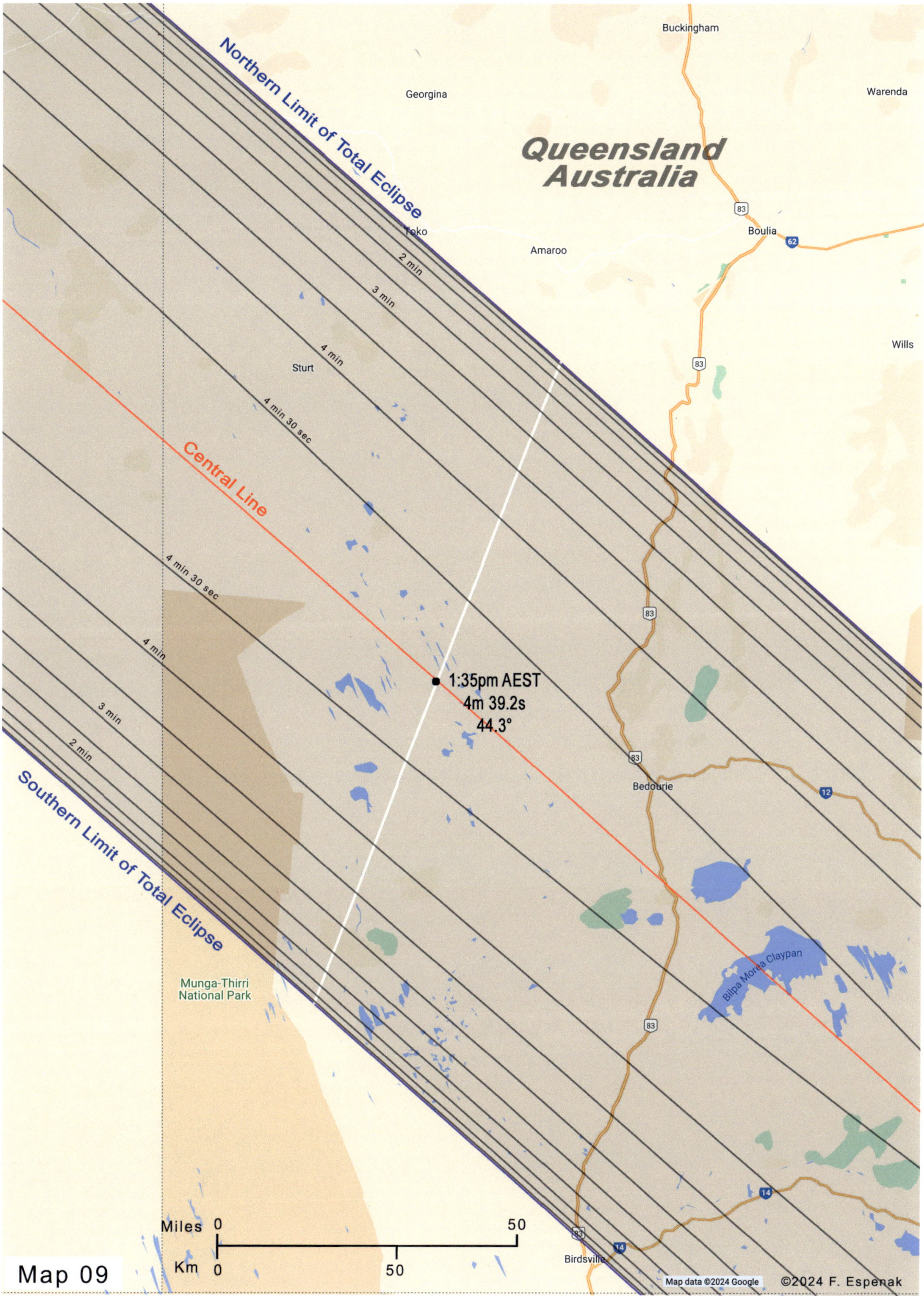
Buckingham
Georgina
Warenda
Queensland
Australia
Northern Limit of Total Eclipse
Toko
Amaroo
Boulia
83
62
2 min
3 min
4 min
Sturt
Wills
4 min 30 sec
Central Line
4 min 30 sec
4 min
1:35pm AEST
4m 39.2s
44.3°
3 min
2 min
Bedourie
12
Southern Limit of Total Eclipse
Bilpa Morea Claypan
Munga-Thirri
National Park
14
Birdsville
Miles 0 50
Km 0 50
Map 09
Map data ©2024 Google
©2024 F. Espenak

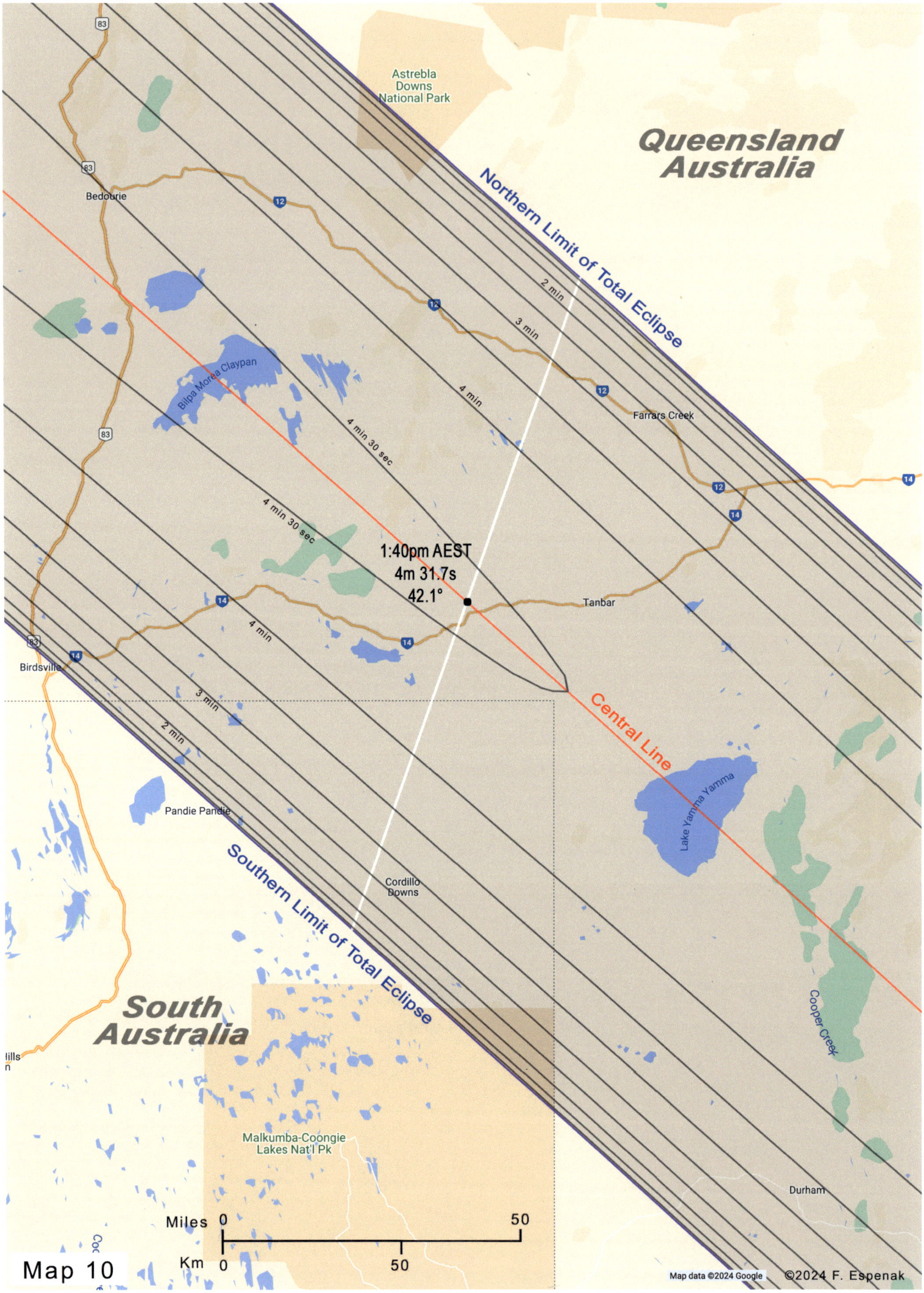
Queensland Australia
South Australia
Northern Limit of Total Eclipse
Southern Limit of Total Eclipse
Central Line
1:40pm AEST
4m 31.7s
42.1°
2 min
3 min
4 min
4 min 30 sec
Astrebla Downs National Park
Malkumba-Coongie Lakes Nat'l Pk
Bedourie
Birdsville
Farrars Creek
Tanbar
Pandie Pandie
Cordillo Downs
Durham
Bilpa Morea Claypan
Lake Yamma Yamma
Cooper Creek
Miles 0 50
Km 0 50
Map 10
Map data ©2024 Google
©2024 F. Espenak

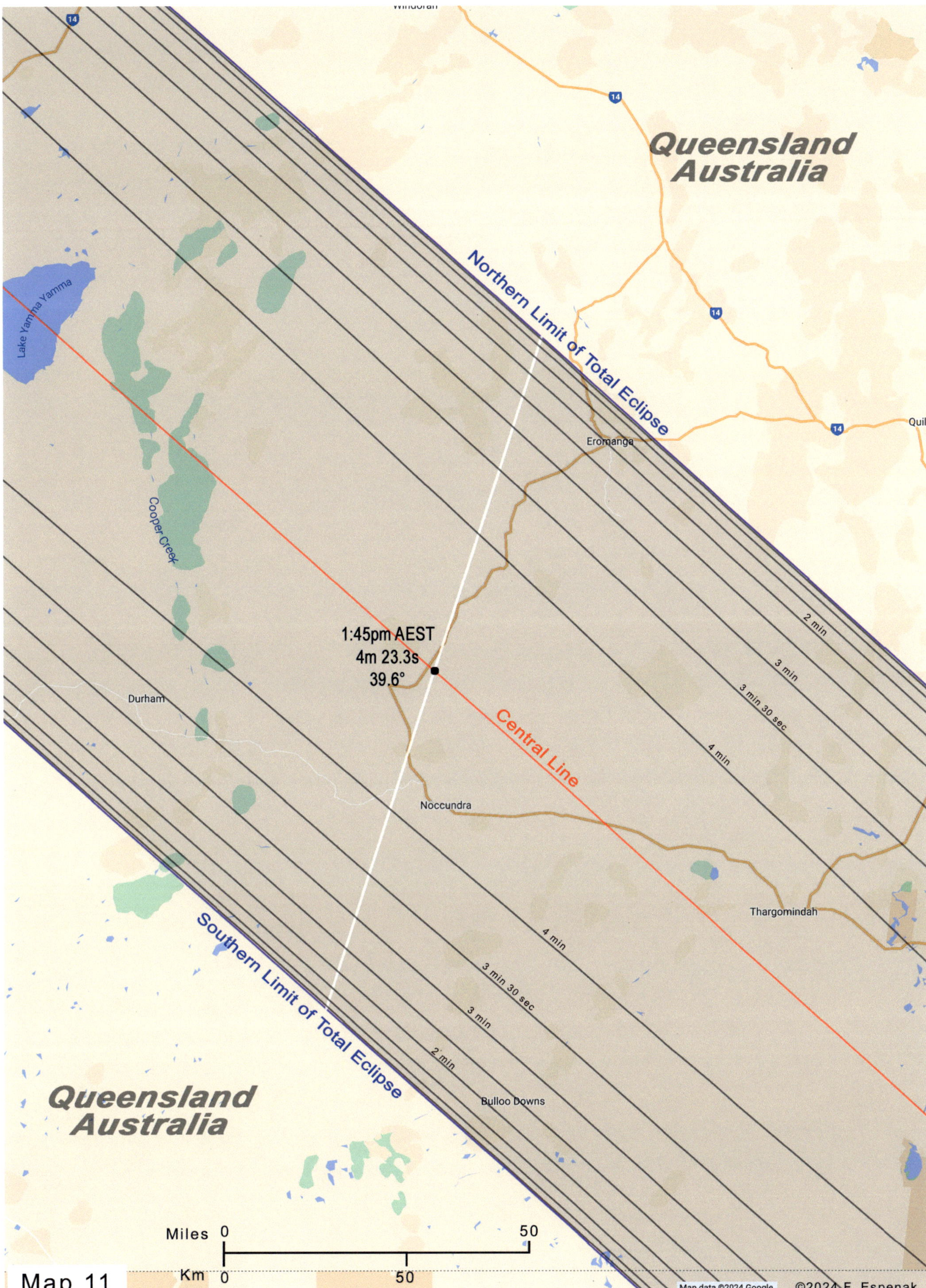
Queensland
Australia
Northern Limit of Total Eclipse
Lake Yamma Yamma
Cooper Creek
Eromanga
Quilpi
1:45pm AEST
4m 23.3s
39.6°
Durham
Central Line
2 min
3 min
3 min 30 sec
4 min
Noccundra
Thargomindah
Southern Limit of Total Eclipse
Bulloo Downs
Queensland
Australia
Miles 0 50
Km 0 50
Map 11
Map data ©2024 Google
©2024 F. Espenak

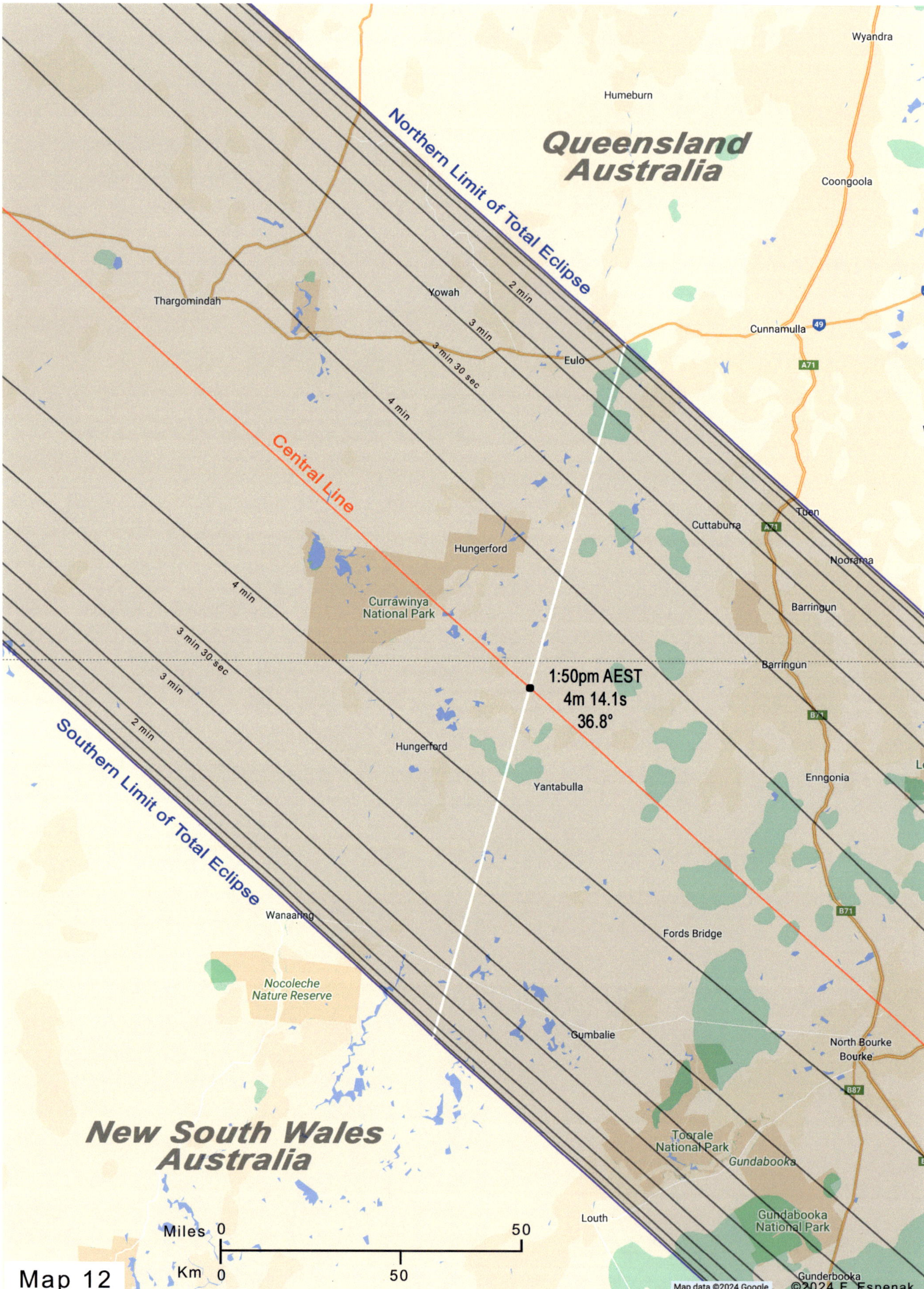
Queensland
Australia
Northern Limit of Total Eclipse
Central Line
Southern Limit of Total Eclipse
New South Wales
Australia
2 min
3 min
3 min 30 sec
4 min
1:50pm AEST
4m 14.1s
36.8°
Wyandra
Humeburn
Coongoola
Thargomindah
Yowah
Cunnamulla
Eulo
Cuttaburra
Tuen
Noorama
Hungerford
Currawinya
National Park
Barringun
Hungerford
Yantabulla
Enngonia
Wanaaring
Fords Bridge
Nocoleche
Nature Reserve
Gumbalie
North Bourke
Bourke
Toorale
National Park
Gundabooka
Gundabooka
National Park
Louth
Gunderbooka
Miles 0 50
Km 0 50
Map 12
Map data ©2024 Google
©2024 F. Espenak

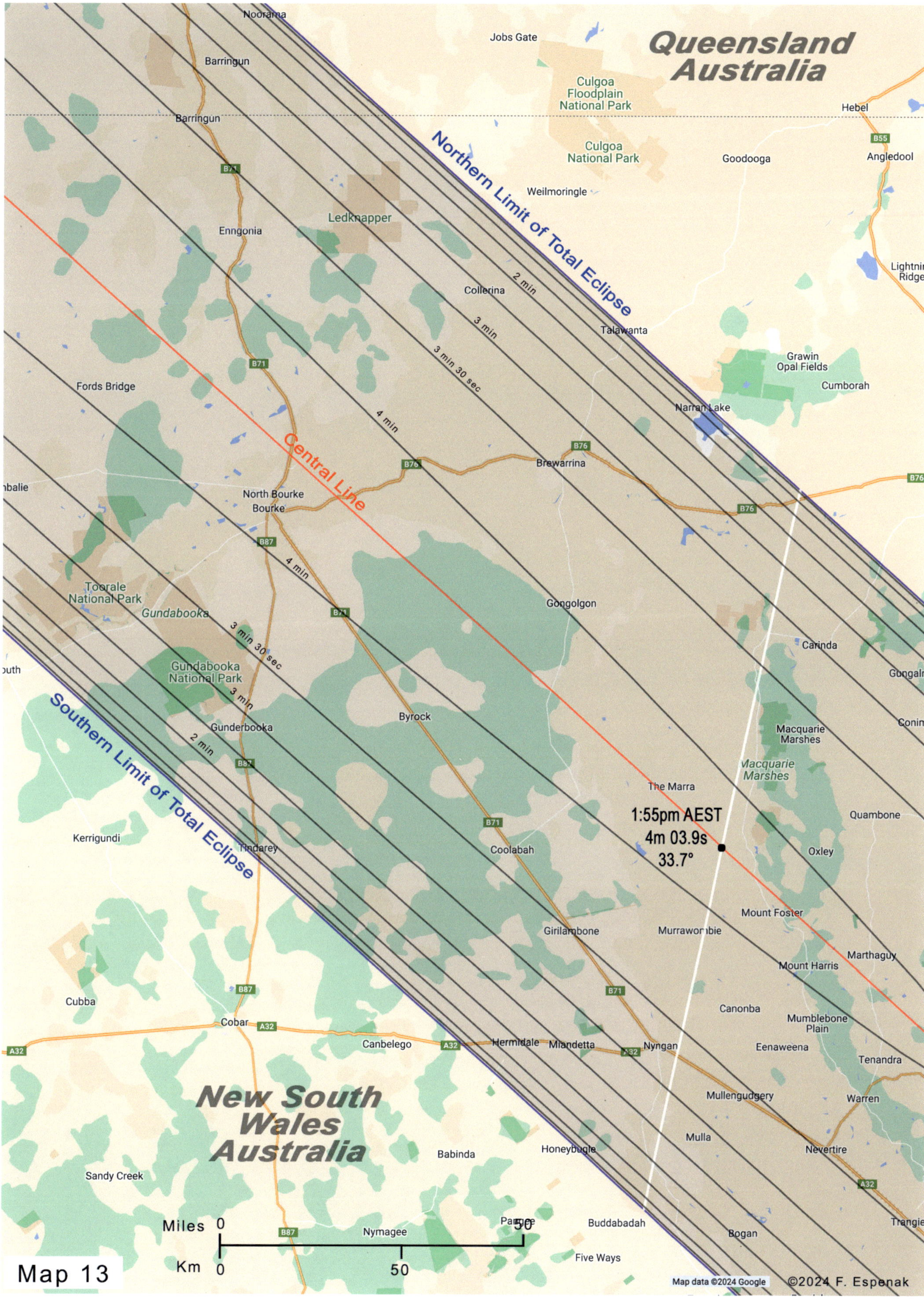
Queensland Australia
New South Wales Australia
Northern Limit of Total Eclipse
Southern Limit of Total Eclipse
Central Line
2 min
3 min
3 min 30 sec
4 min
1:55pm AEST
4m 03.9s
33.7°
Noorama
Jobs Gate
Barringun
Culgoa Floodplain National Park
Culgoa National Park
Hebel
Goodooga
Angledool
Weilmoringle
Ledknapper
Enngonia
Lightning Ridge
Collerina
Talawanta
Grawin Opal Fields
Cumborah
Fords Bridge
Narran Lake
Brewarrina
North Bourke
Bourke
Toorale National Park
Gundabooka
Gongolgon
Carinda
Gundabooka National Park
Gunderbooka
Byrock
Macquarie Marshes
The Marra
Quambone
Kerrigundi
Tindarey
Coolabah
Oxley
Mount Foster
Girilambone
Murrawombie
Marthaguy
Mount Harris
Canonba
Mumblebone Plain
Cubba
Cobar
Canbelego
Hermidale
Miandetta
Nyngan
Eenaweena
Tenandra
Mullengudgery
Warren
Mulla
Nevertire
Babinda
Honeybugle
Sandy Creek
Buddabadah
Trangie
Bogan
Nymagee
Five Ways
Miles 0 50
Km 0 50
Map 13
Map data ©2024 Google
©2024 F. Espenak

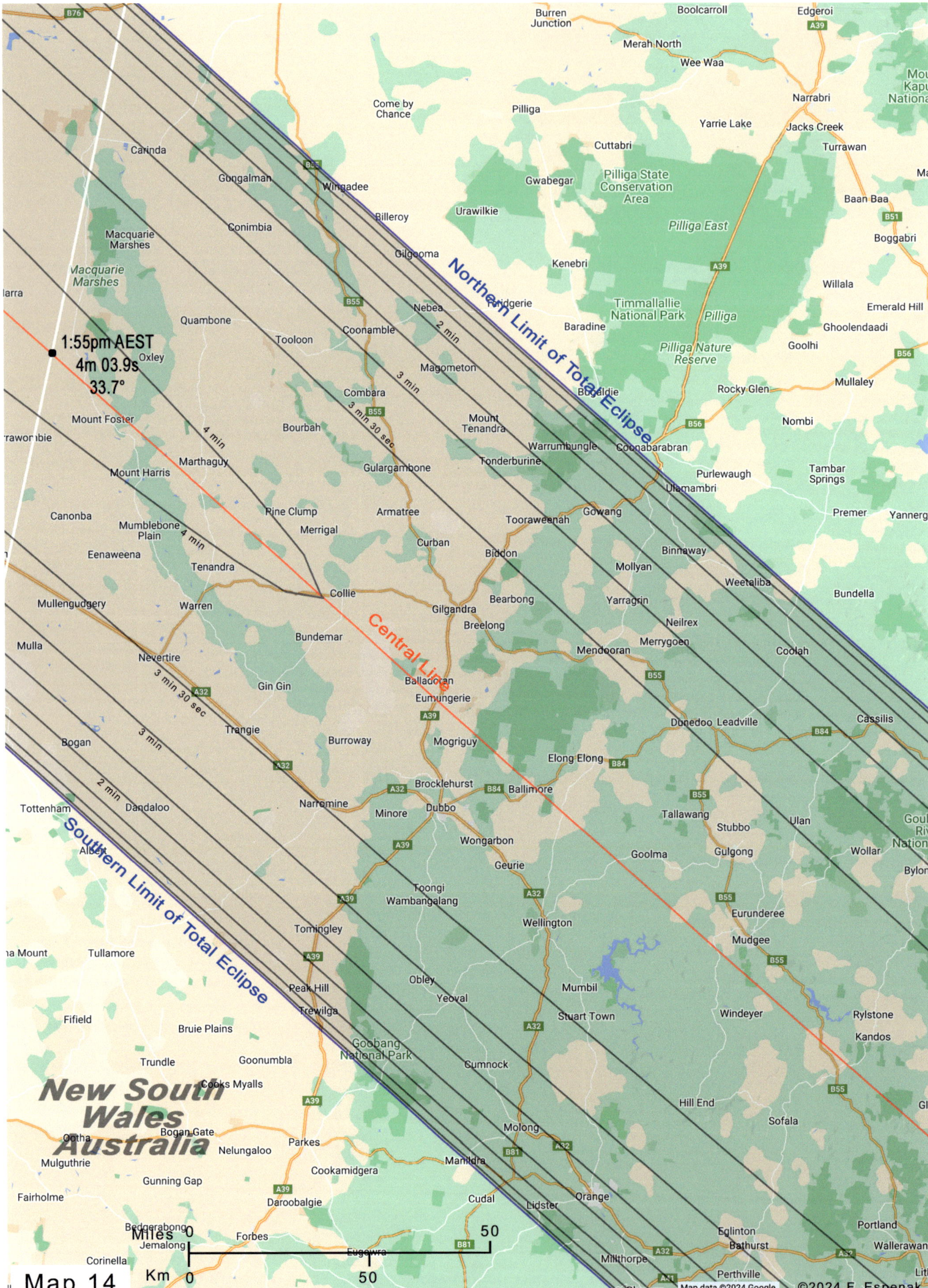
Northern Limit of Total Eclipse
Southern Limit of Total Eclipse
Central Line
1:55pm AEST
4m 03.9s
33.7°
2 min
3 min
3 min 30 sec
4 min
New South Wales Australia
Map 14
Miles 0 50
Km 0 50
Map data ©2024 Google
©2024 F. Espenak
Burren Junction
Boolcarroll
Edgeroi
Merah North
Wee Waa
Come by Chance
Pilliga
Narrabri
Yarrie Lake
Jacks Creek
Turrawan
Carinda
Cuttabri
Gungalman
Wingadee
Gwabegar
Pilliga State Conservation Area
Baan Baa
Billeroy
Urawilkie
Pilliga East
Boggabri
Conimbia
Macquarie Marshes
Gilgooma
Kenebri
Willala
Emerald Hill
Nebea
Baradine
Timmallallie National Park
Ghoolendaadi
Quambone
Coonamble
Tooloon
Pilliga Nature Reserve
Goolhi
Oxley
Magometon
Mullaley
Combara
Rocky Glen
Mount Foster
Bourbah
Mount Tenandra
Nombi
Warrumbungle
Coonabarabran
Marthaguy
Gulargambone
Tonderburine
Mount Harris
Purlewaugh
Tambar Springs
Ulamambri
Canonba
Pine Clump
Armatree
Gowang
Premer
Yannergee
Mumblebone Plain
Merrigal
Tooraweenah
Eenaweena
Curban
Binnaway
Biddon
Tenandra
Mollyan
Weetaliba
Bundella
Collie
Mullengudgery
Warren
Bearbong
Yarragrin
Gilgandra
Breelong
Neilrex
Bundemar
Mendooran
Merrygoen
Mulla
Coolah
Nevertire
Balladoran
Gin Gin
Eumungerie
Trangie
Dunedoo
Leadville
Cassilis
Bogan
Burroway
Mogriguy
Elong Elong
Brocklehurst
Ballimore
Tottenham
Dandaloo
Narromine
Minore
Dubbo
Tallawang
Stubbo
Ulan
Albert
Wongarbon
Goolma
Gulgong
Wollar
Geurie
Bylong
Toongi
Wambangalang
Eurunderee
Tomingley
Wellington
Mudgee
Tullamore
Obley
Peak Hill
Mumbil
Yeoval
Windeyer
Rylstone
Trewilga
Stuart Town
Kandos
Fifield
Bruie Plains
Goobang National Park
Trundle
Goonumbla
Cumnock
Cooks Myalls
Hill End
Sofala
Molong
Ootha
Bogan Gate
Parkes
Nelungaloo
Manildra
Mulgathrie
Cookamidgera
Gunning Gap
Orange
Fairholme
Daroobalgie
Cudal
Lidster
Portland
Bedgerabong
Jemalong
Forbes
Eugowra
Eglinton
Bathurst
Wallerawang
Corinella
Millthorpe
Perthville
Blayney
O'Connell
Garema

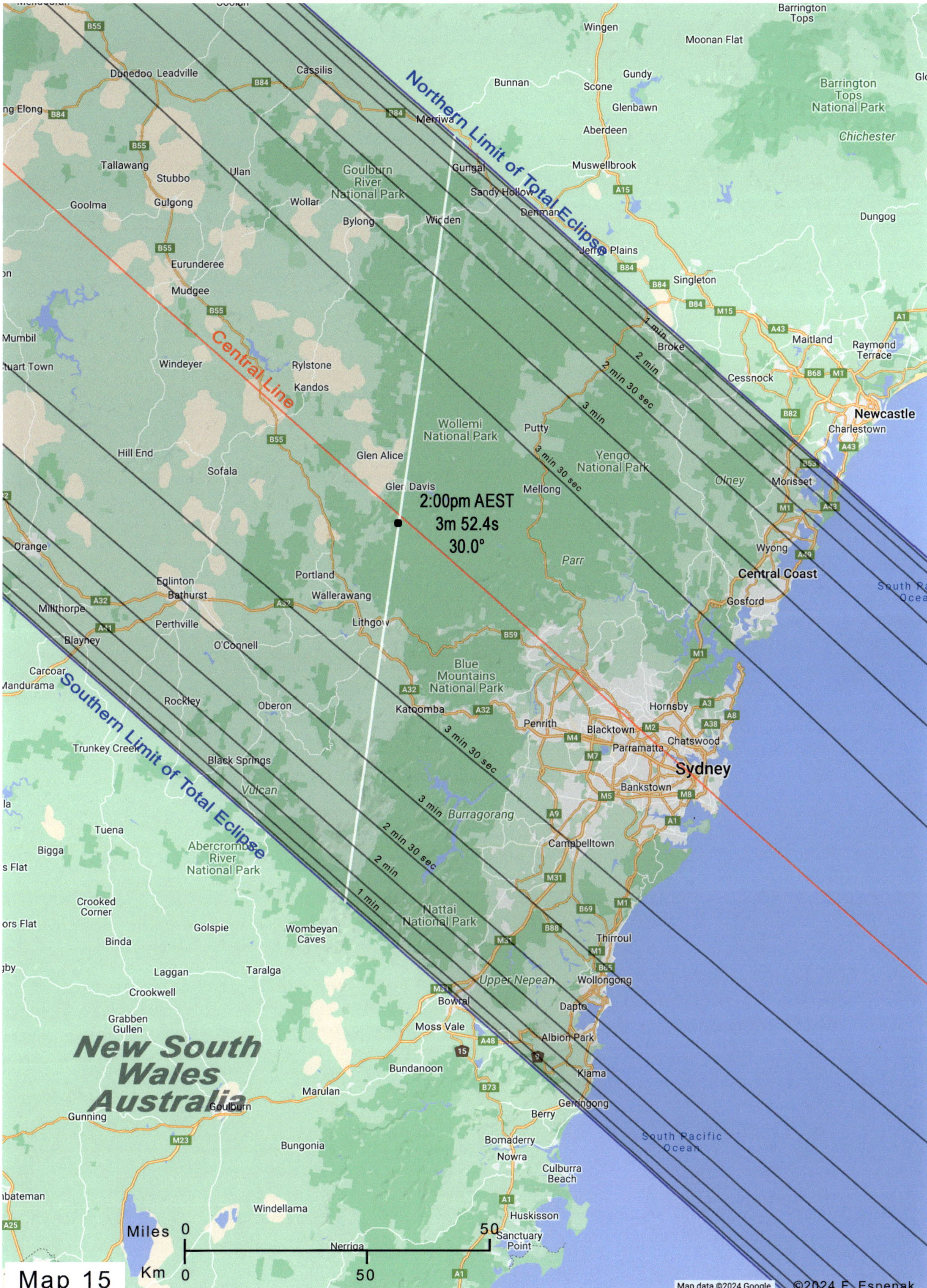
Northern Limit of Total Eclipse
Southern Limit of Total Eclipse
Central Line
2:00pm AEST
3m 52.4s
30.0°
1 min
2 min
2 min 30 sec
3 min
3 min 30 sec
New South Wales Australia
Sydney
Newcastle
Central Coast
Wollemi National Park
Blue Mountains National Park
Yengo National Park
Goulburn River National Park
Abercrombie River National Park
Nattai National Park
Barrington Tops National Park
South Pacific Ocean
Miles 0 50
Km 0 50
Map 15
Map data ©2024 Google
©2024 F. Espenak

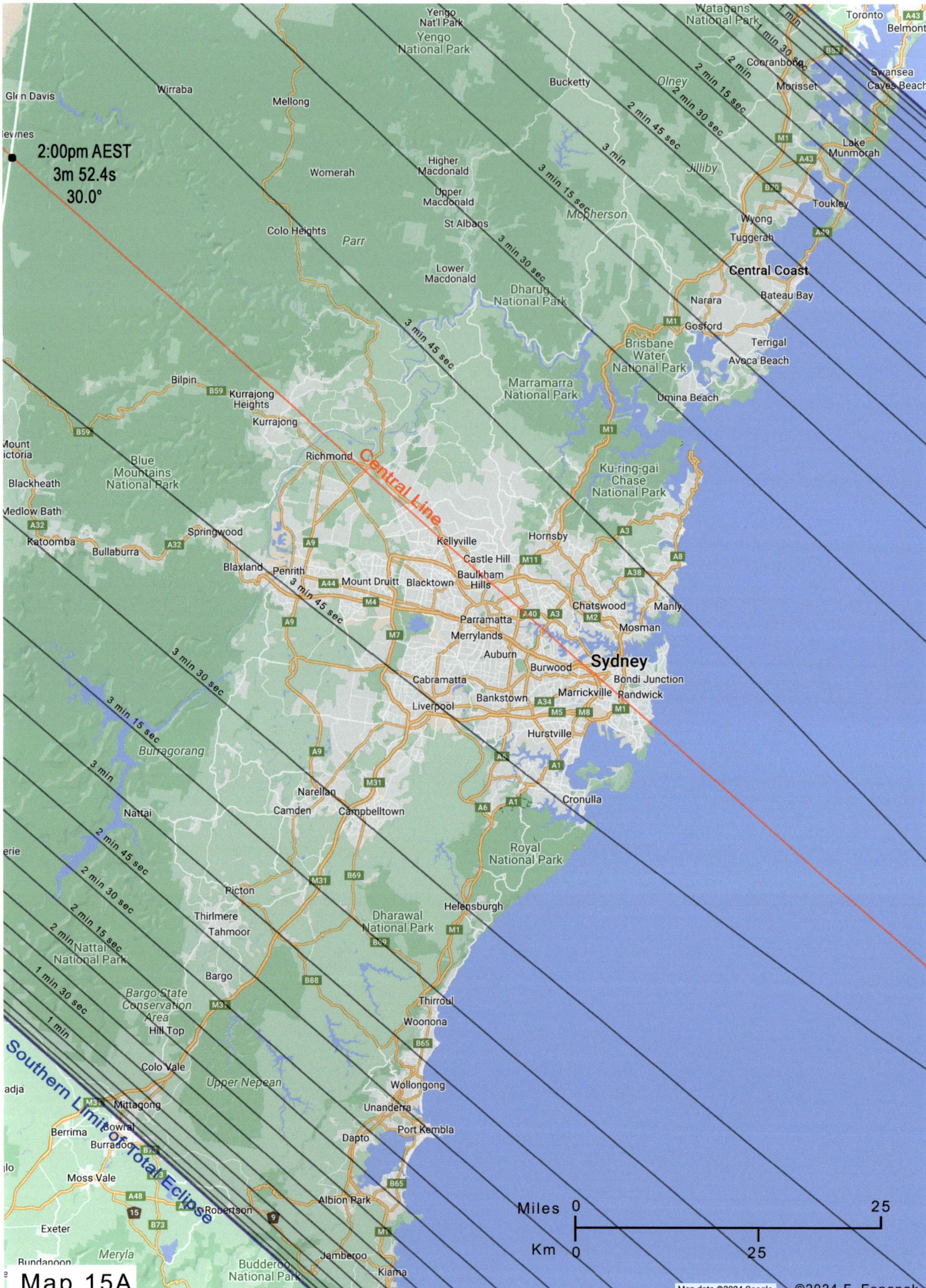
2:00pm AEST
3m 52.4s
30.0°
Central Line
Southern Limit of Total Eclipse
1 min
1 min 30 sec
2 min
2 min 15 sec
2 min 30 sec
2 min 45 sec
3 min
3 min 15 sec
3 min 30 sec
3 min 45 sec
Sydney
Central Coast
Miles 0 25
Km 0 25
Map data ©2024 Google
©2024 F. Espenak
Map 15A

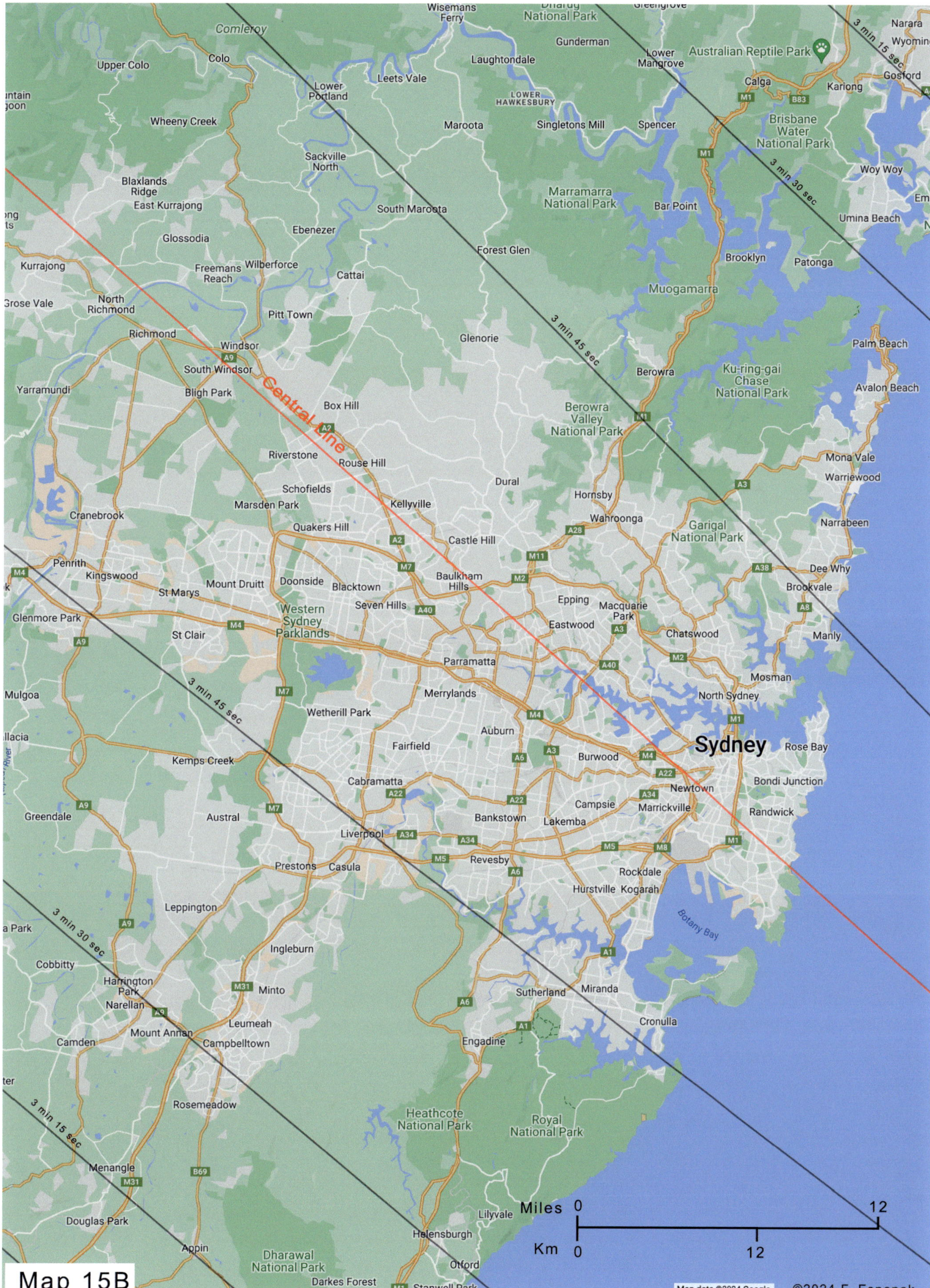
Map 15B
Central Line
3 min 45 sec
3 min 45 sec
3 min 30 sec
3 min 30 sec
3 min 15 sec
3 min 15 sec
Sydney
Wisemans Ferry
Comleroy
Upper Colo
Colo
Wheeny Creek
Lower Portland
Leets Vale
Laughtondale
Gunderman
Lower Mangrove
Australian Reptile Park
Narara
Wyoming
Gosford
Calga
Kariong
LOWER HAWKESBURY
Maroota
Singletons Mill
Spencer
Brisbane Water National Park
Sackville North
Blaxlands Ridge
East Kurrajong
Marramarra National Park
Bar Point
Woy Woy
South Maroota
Umina Beach
Glossodia
Ebenezer
Forest Glen
Kurrajong
Freemans Reach
Wilberforce
Cattai
Brooklyn
Patonga
Grose Vale
North Richmond
Pitt Town
Muogamarra
Richmond
Windsor
Glenorie
Palm Beach
South Windsor
Berowra
Ku-ring-gai Chase National Park
Yarramundi
Bligh Park
Box Hill
Berowra Valley National Park
Avalon Beach
Riverstone
Rouse Hill
Mona Vale
Warriewood
Dural
Schofields
Hornsby
Marsden Park
Kellyville
Cranebrook
Wahroonga
Garigal National Park
Narrabeen
Quakers Hill
Castle Hill
Penrith
Kingswood
Mount Druitt
Doonside
Blacktown
Baulkham Hills
Dee Why
St Marys
Brookvale
Epping
Macquarie Park
Seven Hills
Glenmore Park
Western Sydney Parklands
Eastwood
St Clair
Chatswood
Manly
Parramatta
Mosman
Mulgoa
Merrylands
North Sydney
Wetherill Park
Auburn
Fairfield
Rose Bay
Burwood
Kemps Creek
Cabramatta
Newtown
Bondi Junction
Campsie
Marrickville
Greendale
Austral
Bankstown
Lakemba
Randwick
Liverpool
Revesby
Prestons
Casula
Rockdale
Hurstville
Kogarah
Leppington
Botany Bay
Ingleburn
Cobbitty
Harrington Park
Minto
Narellan
Sutherland
Miranda
Leumeah
Cronulla
Camden
Mount Annan
Campbelltown
Engadine
Rosemeadow
Heathcote National Park
Royal National Park
Menangle
Lilyvale
Douglas Park
Helensburgh
Appin
Dharawal National Park
Otford
Darkes Forest
Stanwell Park
Miles 0 12
Km 0 12
Map data ©2024 Google
©2024 F. Espenak

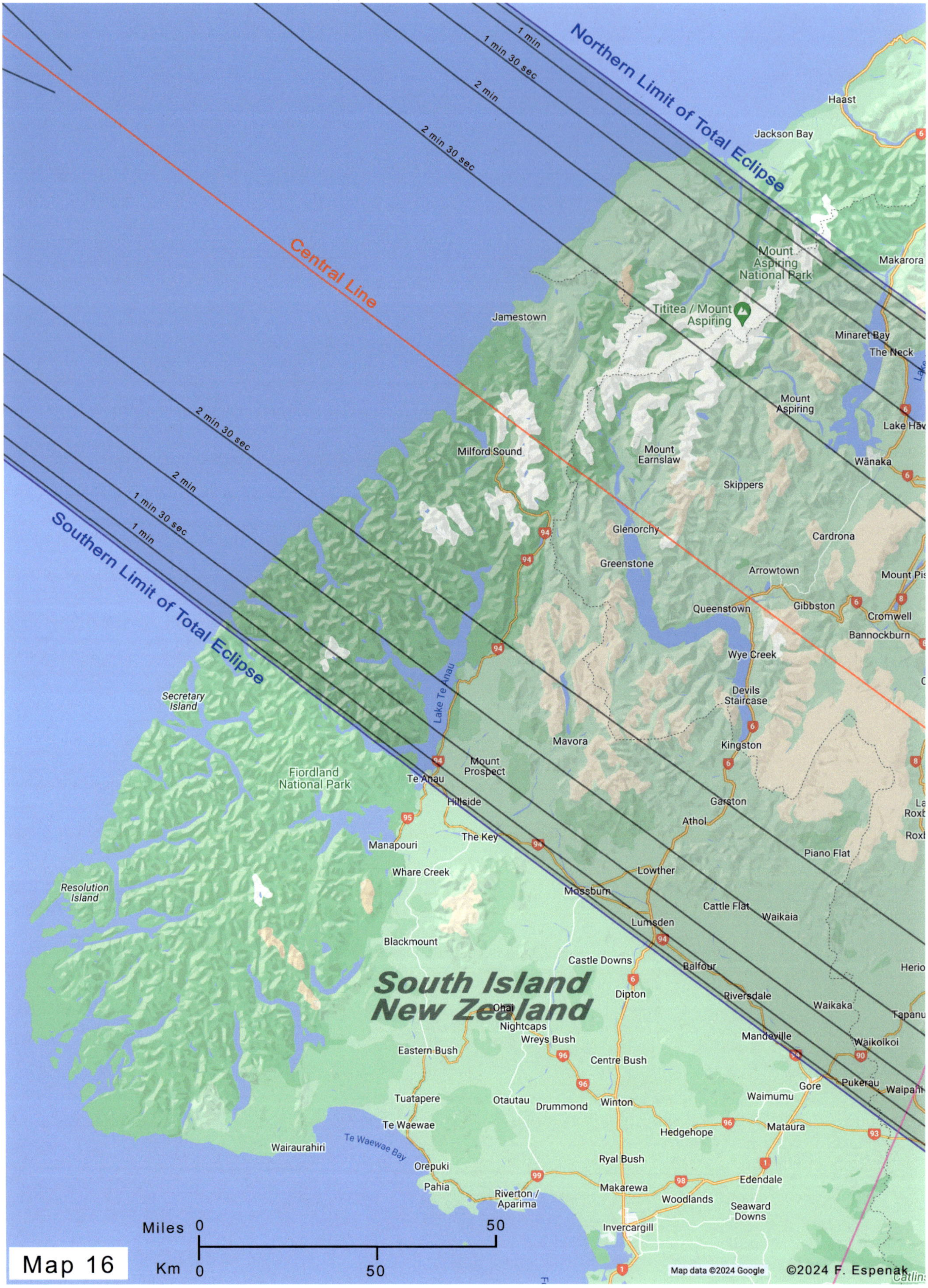
Northern Limit of Total Eclipse
1 min
1 min 30 sec
2 min
2 min 30 sec
Central Line
Southern Limit of Total Eclipse
South Island
New Zealand
Haast
Jackson Bay
Mount Aspiring National Park
Tititea / Mount Aspiring
Makarora
Jamestown
Minaret Bay
The Neck
Mount Aspiring
Milford Sound
Mount Earnslaw
Wānaka
Skippers
Glenorchy
Cardrona
Greenstone
Arrowtown
Queenstown
Gibbston
Cromwell
Bannockburn
Wye Creek
Devils Staircase
Lake Te Anau
Secretary Island
Mavora
Kingston
Te Anau
Mount Prospect
Fiordland National Park
Hillside
Garston
Athol
The Key
Manapouri
Piano Flat
Lowther
Whare Creek
Resolution Island
Mossburn
Cattle Flat
Waikaia
Lumsden
Blackmount
Castle Downs
Balfour
Dipton
Riversdale
Waikaka
Ohai
Nightcaps
Wreys Bush
Mandeville
Waikoikoi
Eastern Bush
Centre Bush
Gore
Pukerau
Tuatapere
Otautau
Drummond
Winton
Waimumu
Te Waewae
Hedgehope
Mataura
Wairaurahiri
Te Waewae Bay
Orepuki
Ryal Bush
Edendale
Pahia
Riverton / Aparima
Makarewa
Woodlands
Seaward Downs
Invercargill
Miles 0 50
Km 0 50
Map 16
Map data ©2024 Google
©2024 F. Espenak

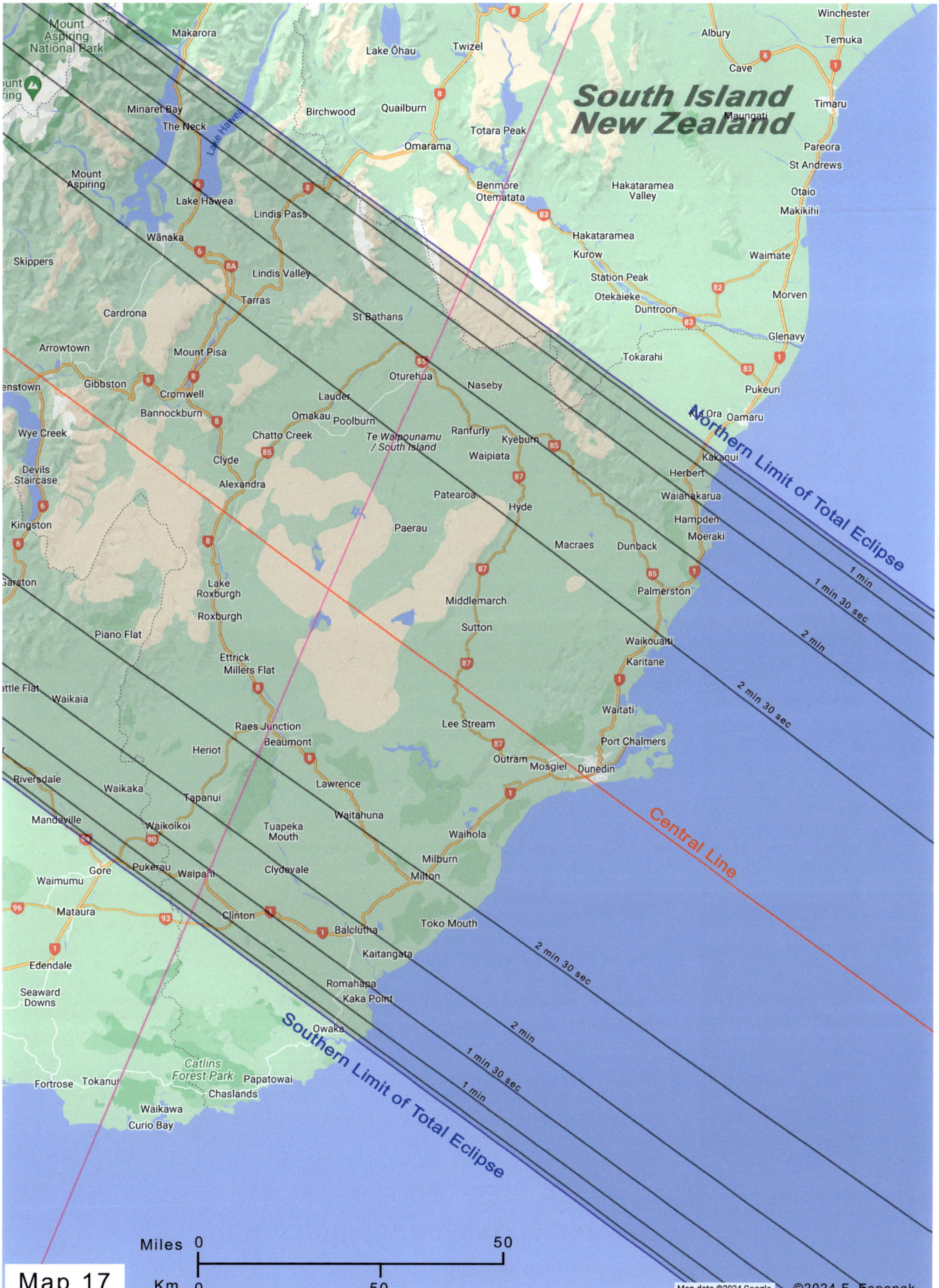
South Island
New Zealand
Northern Limit of Total Eclipse
Southern Limit of Total Eclipse
Central Line
1 min
1 min 30 sec
2 min
2 min 30 sec
2 min 30 sec
2 min
1 min 30 sec
1 min
Mount Aspiring National Park
Makarora
Lake Ōhau
Twizel
Albury
Winchester
Temuka
Cave
Timaru
Maungati
Minaret Bay
The Neck
Lake Hawea
Birchwood
Quailburn
Totara Peak
Omarama
Pareora
St Andrews
Mount Aspiring
Lake Hāwea
Lindis Pass
Benmore
Otematata
Hakataramea Valley
Otaio
Makikihi
Wānaka
Hakataramea
Kurow
Waimate
Skippers
Lindis Valley
Station Peak
Otekaieke
Morven
Tarras
Duntroon
Cardrona
St Bathans
Glenavy
Arrowtown
Mount Pisa
Tokarahi
Gibbston
Oturehua
Naseby
Pukeuri
Cromwell
Lauder
Ma Ora
Oamaru
Bannockburn
Omakau
Poolburn
Wye Creek
Chatto Creek
Te Waipounamu / South Island
Ranfurly
Kyeburn
Kakanui
Waipiata
Herbert
Clyde
Devils Staircase
Alexandra
Patearoa
Waianakarua
Hyde
Hampden
Kingston
Paerau
Moeraki
Macraes
Dunback
Lake Roxburgh
Palmerston
Middlemarch
Roxburgh
Sutton
Piano Flat
Waikouaiti
Karitane
Ettrick
Millers Flat
Waikaia
Waitati
Raes Junction
Lee Stream
Beaumont
Port Chalmers
Heriot
Outram
Mosgiel
Dunedin
Riversdale
Waikaka
Lawrence
Tapanui
Mandeville
Waikoikoi
Waitahuna
Tuapeka Mouth
Waihola
Gore
Pukerau
Waipahi
Clydevale
Milburn
Waimumu
Milton
Mataura
Clinton
Toko Mouth
Balclutha
Kaitangata
Edendale
Seaward Downs
Romahapa
Kaka Point
Owaka
Catlins Forest Park
Fortrose
Tokanui
Papatowai
Chaslands
Waikawa
Curio Bay
Miles 0 50
Km 0 50
Map data ©2024 Google
©2024 F. Espenak

Map 17

Eclipse Guides from AstroPixels Publishing

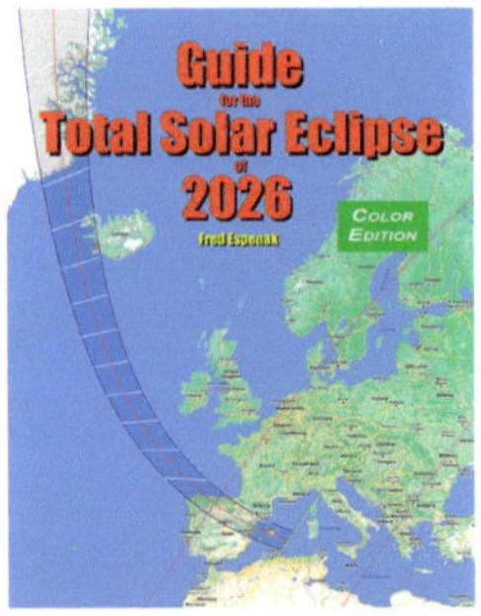

Guide for the Total Solar Eclipse of 2026

The total solar eclipse of August 12, 2026 will be visible from a path across northern Russia, the Arctic, Greenland, Iceland, and Spain. The **Guide for the Total Solar Eclipse of 2026** contains 14 full-page maps of the path. The large scale shows roads, towns and cities, rivers and lakes.

Tables for hundreds of cities provide times each eclipse phase of the along with the eclipse magnitude, duration, and the Sun's altitude. A climatology study identifies areas along the eclipse path where the highest probability of favorable weather may be found.

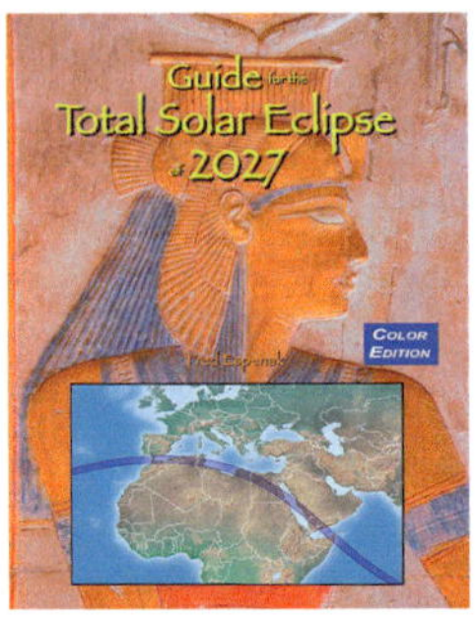

Guide for the Total Solar Eclipse of 2027

The total solar eclipse of August 2, 2027 will be visible from a path across Spain, Morocco, Algeria, Tunisia, Libya, Egypt, Sudan, Saudia Arabia, Yemen, and Somalia. The **Guide for the Total Solar Eclipse of 2027** contains 20 full-page maps of the path.

Tables for hundreds of cities provide times for each eclipse phase along with the eclipse magnitude, duration and the Sun's altitude. Cloud statistics identify areas where the highest probability of clear sky may be found.

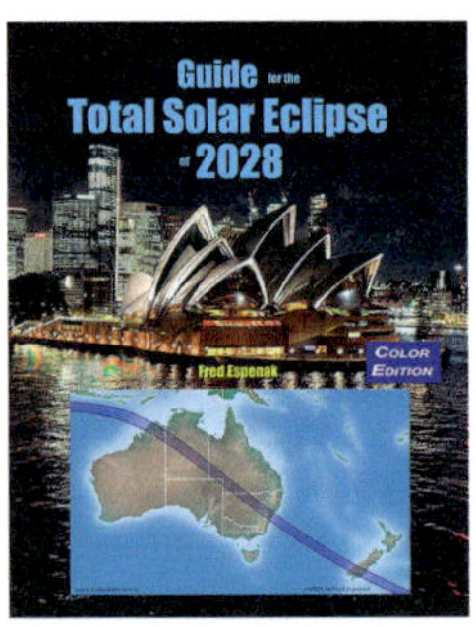

Guide for the Total Solar Eclipse of 2028

The total solar eclipse of July 22, 2028 will be visible from a path across Australia and New Zealand. The **Guide for the Total Solar Eclipse of 2028** contains 19 full-page maps of the path of totality.

Tables for over one hundred cities provide times for each eclipse phase with the duration of totality and the Sun's altitude. Cloud statistics identify areas where the highest probability of clear sky may be found. A preview is given for total eclipses through 2039.

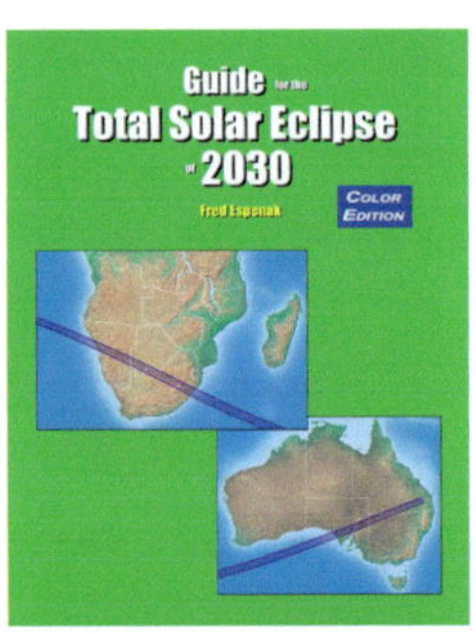

Guide for the Total Solar Eclipse of 2030

The total solar eclipse of November 25, 2030 will be visible from a path across Africa and Australia. The **Guide for the Total Solar Eclipse of 2028** contains 17 full-page maps of the path of totality.

Tables for over one hundred cities provide times for each eclipse phase with the duration of totality and the Sun's altitude. Cloud statistics identify areas where the highest probability of clear sky may be found. A preview is given for total eclipses through 2039.

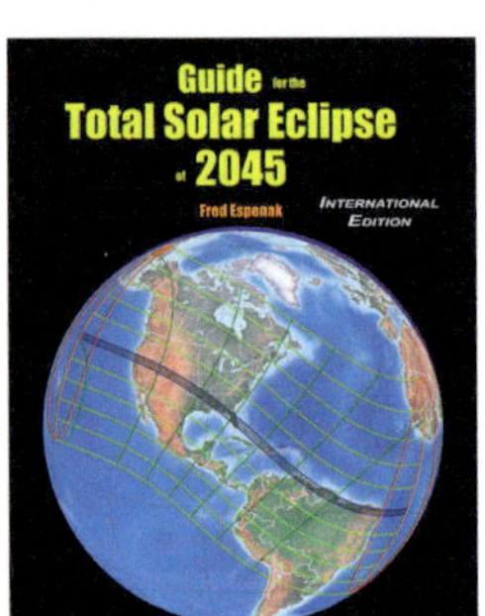

Guide for the Total Solar Eclipse of 2045

The total solar eclipse of August 12, 2045 will be visible from a path across the USA, the Caribbean, and South America. The **Guide for the Total Solar Eclipse of 2045** contains 37 full-page maps of the path of totality.

Tables for hundreds cities provide times for each eclipse phase with the duration of totality, eclipse magnitude, and the Sun's altitude. Cloud statistics identify areas where the highest probability of clear sky may be found.

For more information on each *Guide*, including sample pages of each, visit:

astropixels.com/pubs/guides.html

The *Guides* may be ordered from ***Amazon***

AstroPixels Publishing

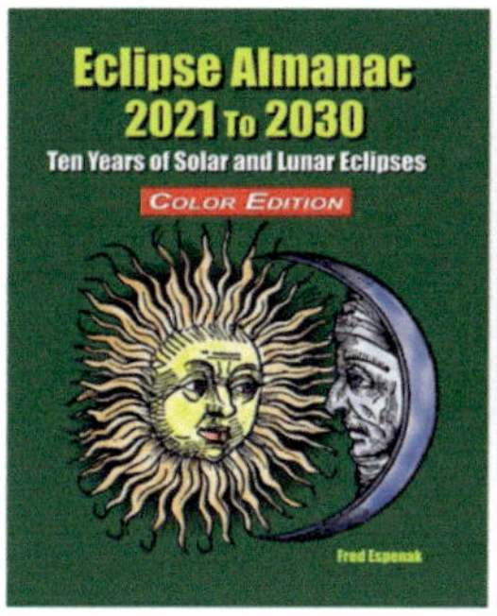

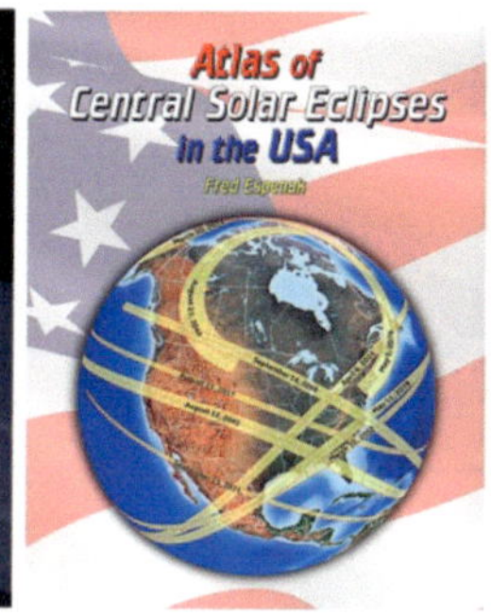

Eclipse Almanac – 2021 to 2030

A concise reference for every eclipse of the Sun and the Moon over a 10-year period. It identifies when and where each event will be seen with maps for each. Additional details about each eclipse are included. Other volumes in this series include the decades 2032 to 2040, 2041 to 2050, 2051 to 2060, and 2061 to 2070.

21st Century Canon of Solar Eclipses

The complete guide to every solar eclipse occurring from 2001 tom 2100 (224 eclipses in all). It includes information and maps for all total, annular, hybrid, and partial eclipses. A special world atlas shows detailed, full page maps of all central eclipse paths (total, annular, and hybrid).

Thousand Year Canon of Solar Eclipses — 1501 to 2500

Contains maps and data for each of the 2,389 solar eclipses occurring over the ten-century period centered on the present era. The maps are arranged twelve to a page at an image scale allowing the determination of eclipse visibility from any location on Earth. A comprehensive catalog lists the essential characteristics of each eclipse.

Atlas of Central Solar Eclipses in the USA

When was the last total eclipse through the USA and when is the next? How often do they happen? What total eclipse tracks passed across the USA during the 17th, 18th, and 19th centuries, etc., and what states did they include? And how often is a total solar eclipse visible from each of the 50 states? The Atlas of Central Solar Eclipses in the USA answers all these questions and more, with hundreds of maps and tables.

21st Century Canon of Lunar Eclipses

The complete guide to every lunar eclipse occurring from 2001 tom 2100 (228 eclipses in all). It includes information and maps for all total, partial, and penumbral eclipses. The predictions use a new model for Earth's elliptical shadows.

Thousand Year Canon of Lunar Eclipses — 1501 to 2500

Contains maps and data for each of the 2,424 lunar eclipses occurring over the ten-century period centered on the present era. The maps are arranged twelve to a page at an image scale allowing the determination of eclipse visibility from any location on Earth. A comprehensive catalog lists the essential characteristics of each eclipse.

All books are available in two editions: 1) Black and White, and 2) Color.

For more information including sample pages of each, visit:

astropixels.com/pubs/guides.html

Each *Book* may be ordered from ***Amazon***

Made in the USA
Monee, IL
12 September 2026

62975591R00029